Prepare for Launch

The Astronaut Training Process

Erik Seedhouse

Prepare for Launch

The Astronaut Training Process

Dr Erik Seedhouse, F.B.I.S., As.M.A.
Milton
Ontario
Canada

SPRINGER–PRAXIS BOOKS IN SPACE EXPLORATION
SUBJECT *ADVISORY EDITOR*: John Mason, M.B.E., B.Sc., M.Sc., Ph.D.

ISBN 978-1-4419-1349-4 Springer Berlin Heidelberg New York

Springer is a part of Springer Science + Business Media (*springer.com*)

Library of Congress Control Number: 2009936077

Apart from any fair dealing for the purposes of research or private study, or criticism or review, as permitted under the Copyright, Designs and Patents Act 1988, this publication may only be reproduced, stored or transmitted, in any form or by any means, with the prior permission in writing of the publishers, or in the case of reprographic reproduction in accordance with the terms of licences issued by the Copyright Licensing Agency. Enquiries concerning reproduction outside those terms should be sent to the publishers.

© Copyright, 2010 Praxis Publishing Ltd., Chichester, UK

The use of general descriptive names, registered names, trademarks, etc. in this publication does not imply, even in the absence of a specific statement, that such names are exempt from the relevant protective laws and regulations and therefore free for general use.

Cover design: Jim Wilkie
Project copy editor: Christine Cressy
Typesetting: BookEns, Royston, Herts., UK

Printed in Germany on acid-free paper

Contents

Preface

"I have always wanted to be an astronaut. Every kid growing up during the space race wanted to be an astronaut. Unlike most kids, however, I never vacillated in my career choices. I have a vivid memory of sitting on the lounge floor, watching the first lunar landing on my parents' black and white television set. Watching the images of the *Saturn V* on the launch pad and astronauts bounding across the lunar surface, I resolved to become an astronaut. Only 5 years old at the time, the initial appeal for me of becoming an astronaut was riding a rocket and walking on the Moon.

As I learned more about what was required to be selected as an astronaut, I focused increasingly on achieving my dream. In 1987, knowing many astronauts had a military background, I joined the 2nd Battalion of the Parachute Regiment. Forty-three recruits started the 27-week basic training to become a 'Para'. Just three of us were successful. I was one of the three.

During my time in the Paras, I was trained by the Special Air Service (SAS) in jungle and desert warfare, made night jumps from Chinook helicopters, and jumped out of a Hercules C-130 more times than I can remember. I also endured annual 80-km marches with a 20-kg pack on my back and acquired advanced survival techniques in the jungles of Belize and the desert of northern Cyprus. During operational deployments, I gained extensive experience operating in the cohesive environment of a small team, while under considerable and prolonged stress. I enjoyed the training immensely, especially knowing it provided excellent experience for becoming an astronaut.

My first postgraduate degree was a Master's in Medical Science at Sheffield University. Rather than choose one of the research options offered by the course, I decided to pay Dr. David Grundy a visit. David was Director of the Institute for Space Biomedicine. Together, we formulated a space life-sciences research proposal focusing on head-down-tilt, the outcome of which was an article published in the *Physiologist* in 1991. While at Sheffield, I began to focus more on sports, with the goal of becoming a professional athlete.

On completing my Master's degree, I moved on to work in a Pulmonary Function

Laboratory and learned how to balance budgets and develop test protocols. Shortly after starting work at the laboratory in March, 1992, I applied to the Canadian Astronaut Program. Since I had not yet accumulated the myriad qualifications required for a competitive application, my objective was to indicate to the CSA my interest in becoming an astronaut, with the intention of being able to submit a more competitive application when the next selection was advertised.

Although I realized the next selection might be many years away, my focus never wavered. My next goal was to experience microgravity and, to that end, I volunteered to be a subject in a study conducted by a colleague of Dr. Grundy. In April, 1995, I participated as a subject in the European Space Agency's 22nd Parabolic Flight Campaign, which provided me with valuable experience in performing equipment set-up, calibration, and monitoring in support of physiological testing. The data collection for the study took place over a 5-day period, flying 30 parabolas daily over the Bay of Biscay, providing me with ample opportunity to experience microgravity.

My increasing involvement in competitive sports culminated in my achieving a key ambition: that of becoming a professional athlete. This goal was accomplished after winning the World Endurance Triathlon Championships in 1995. During my triathlon career, I repeatedly demonstrated my commitment to my corporate sponsors, an enduring resolve, and the mental and physical fortitude enabling me to win what many considered impossible races. Winning world championship races was achieved by adhering to a strict and rigorous training regime and with the motivation to achieve distinction, a trait I consider an asset to any aspiring astronaut. Furthermore, due to my success, I was in regular demand by the media for interviews and gained invaluable experience as a representative for my corporate sponsors.

Shortly after becoming a professional triathlete, I moved to Germany, courtesy of a $50,000 ESA grant. While studying for my Ph.D. at the Institute for Space Medicine, I had the opportunity to acquire a working knowledge of ESA's program of operation. My work at the Institute also provided me with experience related to a variety of space program development practices and extended my ability to interact with external and international scientific counterparts. I pursued my Ph.D. studies simultaneously with my sports career – a combination requiring me to work 70 or more hours a week. Apart from being an ideal place to conduct research, the Institute also provided an outstanding opportunity to accumulate expertise in space program procedures and acquire a working knowledge of space medicine.

Upon retiring as a triathlete in 1999, I pursued my post-doctoral qualification at Simon Fraser University's (SFU) Environmental Physiology Unit (EPU). While there, I envisioned the Extreme Physiology Program (EPP) and initiated field research trips to Mount McKinley and Aconcagua. The EPP was envisaged as a multi-science/technology research and development program in which studies of human physiology in stressful environments would be conducted. These studies included lab experiments (simulated space, high altitude, and deep sea) and field studies (using wireless technologies), the first of which was a simulated 5-day stay onboard the International Space Station. Later (field) studies included ascents of Mt.

McKinley and Aconcagua to investigate the differences in physiological responses between athletes and non-athletes. As a Co-Director of the EPP, I directed, implemented, managed, and monitored research programs, and developed the priorities and objectives for research projects within the program.

In 2005, I was recruited by Bigelow Aerospace, Las Vegas, as an Astronaut Training Consultant. During my time at Bigelow Aerospace, I designed, wrote, and edited the *Spaceflight Participants Training Manual*, designed medical algorithms for spaceflight medical emergencies, and developed instructional materials. I also designed, wrote, and edited the *Chief Medical Officer's Manual* for spaceflight participant selection and designed and edited astronaut certification standards for spaceflight participants.

In 2007, I joined the Aerospace Group at the Defence and Research Development Canada (DRDC) in Toronto, to work as a program manager overseeing specialized research development projects.

Curiosity has always gotten the better of me. I enjoy scuba-diving, climbing mountains, writing books, racing Ironman triathlons, flying my Cessna, and writing and directing movies. My quirk is that I prefer technical manuals to television and I hate sitting still. I cannot resist new activities and enjoy squeezing as many projects into my life as possible. I consider the ability to display competence, self-reliance, and the ability to work effectively even under the most challenging conditions as traits required by an astronaut and I believe I possess those qualities. Having varied experiences and adapting well to their demands is a desirable characteristic for an astronaut and I consider my background supports this requirement.

My varied work experience has provided me with the opportunity to develop excellent communication skills, both written and verbal. I have contributed extensively to *Spaceflight* magazine, been interviewed on numerous occasions, published four books, and have worked as a motivational speaker. As an expedition leader, I utilized logistical skills to achieve successful outcomes in dynamic and challenging environments. I consider these and the other aforementioned skills to be key to my effective contribution to the CSA and I am confident in my ability to respond to the challenges of being a Canadian astronaut.

As a potential astronaut, I have demonstrated a willingness to accept physical hazards, a capacity to tolerate rigorous and severe environmental conditions, and the ability to react assertively under conditions of duress. These abilities have been demonstrated during parachute jumps, mountaineering, diving, triathlon, microgravity research, and as an instructor during high-altitude indoctrination training.

As an astronaut, I believe I will be an asset to the CSA. My varied scientific experience, combined with my athletic career and military training, have provided me with a unique background that has inculcated a calm, methodical, and cautioned approach to problem solving, whether in a laboratory or under conditions of prolonged physical and mental duress. I am easy to work with and integrate easily within a team and I believe I can provide the same support, encouragement, and honesty to fellow co-workers and astronauts.

Throughout my career, I have maintained an enduring, positive commitment to manned spaceflight, my life's one overriding passion. I have been reading about and studying spaceflight since that day I watched the Moon landings. I want to join the Canadian Astronaut Team because it is the best occupation to fully utilize my professional skills and experience, and because it is a profession in which I could contribute the most to the CSA in its execution of Canada's space policy. I would be privileged to join the Canadian Astronaut Team, who have raised teamwork to a new level, made excellence a given, and challenged conventional thinking with courage and imagination. I am ready and eager to be a part of that team."

One of the first stages of the CSA's 2008 Astronaut Recruitment Campaign required applicants to write an essay explaining why they wanted to be an astronaut. The essay you have just read was the one I submitted. After a year of selection tests, including sea survival, fighting fires, robotics assessment, and a myriad other trials and evaluations, I made it to the final 30. Unfortunately, I didn't make it any further. For those applicants who had dedicated 20 years or more to the goal of becoming an astronaut, the blow was devastating to put it mildly. Unlike NASA, which routinely recruits astronauts, the CSA's campaign was only the third since 1983. For many of those deselected at the final stage who had waited 16 years since the CSA's 1992 campaign, it was the end of the road. The lucky few who worked in the US indicated they would apply for a Green Card with the intent of applying to NASA at the next campaign. Others returned to their jobs in academia, flying jets, or performing research. A small number found themselves recalibrating their life, trying to find another challenge. For example, one of the most highly qualified candidates confided to me he really didn't like his job (he is one of the world's leading emergency physicians and an Everest summiteer) and was considering climbing K2. Some reflected on the high price they had paid in the personal lives (many had burned through more than one marriage) by following a dream demanding sacrifices few can fathom. Then there were candidates such as myself who set their sights on achieving their goal by gaining employment (and hopefully a flight!) with one of the fledgling private space companies such as Virgin Galactic and SpaceX.

The road to becoming an astronaut is one demanding tremendous sense of direction, perspective, resolve, and extraordinary self-confidence. While many have entertained thoughts of becoming an astronaut, few ever embark upon the journey requiring them to accumulate the myriad qualifications essential for a competitive application. Those ultimately selected are already some of the most highly trained humans on the planet, yet the preparation for their new occupation hasn't even begun! Here, in this book, you will learn how these extraordinarily qualified and uniquely gifted people learn to be astronauts and how their unique training prepares them to do the most exciting, rewarding, and fulfilling job on and off the Earth.

Outline of the Chapters

There have been several books written by astronauts describing the challenges of training and preparing for spaceflight, but most of these have focused on time in orbit. Thanks to first-person accounts such as *Sky Walking* by Tom Jones, and

Riding Rockets by Mike Mullane, armchair spacefarers have the opportunity to gain an insight into the orbital adventures of astronauts. The objective of this book, however, is to describe the nuts and bolts of astronaut training, starting with the application process and finishing with the climactic ride into space.

Section I starts with the application and selection process. As a two-time applicant to become an astronaut with the Canadian Space Agency (CSA), the author reveals what it takes to assemble a competitive application and describes the stringent selection criteria used by space agencies to select future astronauts.

Section II begins with a generic chronology of an astronaut's probationary training at Johnson Space Center (JSC). From spending freezing nights in the Absaroka Mountains learning how to survive on rabbit stew, to pulling Gs in T-38s, the author describes the myriad training elements comprising an astronaut candidate's (ascan) first step towards becoming a fully fledged astronaut. Section II also provides an insight into the world of technical assignments, the seemingly never-ending wait for a flight, and an overview of the types of missions astronauts will be flying over the next two decades. Following a detailed account of the multitude of training elements comprising mission training, the author peers into the future by describing the advantages of hibernation and the challenges of bioethics training.

Before Section III delves into the high-intensity world of pre-launch preparation, the author provides an overview of NASA's new family of launch vehicles and spacecraft that will transport astronauts to the International Space Station (ISS), the Moon and, eventually, Mars. Following a "behind-the-scenes" glimpse of the launch team, the reader is guided through the final 10 weeks of mission training and preparation ultimately leading to launch day.

Acknowledgments

In writing this book, I have been fortunate to have had my wife, Doina, as my proof-reader. Once again, she has applied her considerable skills to make the text as smooth and coherent as possible. Any remaining shortcomings are my responsibility and mine alone.

I am also grateful to the five reviewers who made such positive comments concerning the content of this publication and to Clive Horwood and his team at Praxis for guiding this book through the publication process. The author also gratefully acknowledges Christine Cressy and BookEns, whose attention to detail and patience greatly facilitated the publication of this book. Thanks also to Jim Wilkie for creating the cover of this book.

Once again, no acknowledgment would be complete without special mention of our cats, Jasper and MiniMach, who provided endless welcome distraction and entertainment.

This book is dedicated primarily to my wife, without whom I would never have had the opportunity to pursue my dream.

It is also dedicated to those who helped me along the way, such as Professor David Grundy, Professor Paul Enck, and Parvez Kumar.

Finally, this book is dedicated to Heike, Rolf, Christian, Mark, and dozens of other supremely qualified individuals who, due to myopic political agendas or by not having the right passport, were denied an opportunity to demonstrate their talents as astronauts.

About the author

Erik Seedhouse is an aerospace scientist whose ambition has always been to work as an astronaut. After completing his first degree in Sports Science at Northumbria University, the author joined the legendary 2nd Battalion the Parachute Regiment, the world's most elite airborne regiment. During his time in the "Paras", Erik spent 6 months in Belize, where he was trained in the art of jungle warfare and conducted several border patrols along the Belize–Guatemala border. Later, he spent several months learning the intricacies of desert warfare on the Akamas Range in Cyprus. He made more than 30 jumps from a Hercules C130 aircraft, performed more than 200 abseils from a helicopter, and fired more light anti-tank weapons than he cares to remember!

Upon returning to the comparatively mundane world of academia, the author embarked upon a Master's degree in Medical Science at Sheffield University. He supported his Master's degree studies by winning prize money in 100 km ultradistance running races. Shortly after placing third in the World 100 km Championships in 1992 and setting the North American 100 km record, the author turned to ultradistance triathlon, winning the World Endurance Triathlon Championships in 1995 and 1996. For good measure, he also won the inaugural World Double Ironman Championships in 1995 and the infamous Decatriathlon, the world's longest triathlon, an event requiring competitors to swim 38 km, cycle 1,800 km, and run 422 km. Non-stop!

Returning to academia once again in 1996, Erik pursued his Ph.D. at the German Space Agency's Institute for Space Medicine. While conducting his Ph.D. studies, he still found time to win Ultraman Hawaii and the European Ultraman Championships as well as completing the Race Across America bike race. Due to his success as the world's leading ultradistance triathlete, Erik was featured in dozens of magazines and television interviews. In 1997, *GQ* magazine nominated him as the "Fittest Man in the World".

In 1999, Erik decided it was time to get a real job. He retired from being a professional triathlete and started his post-doctoral studies at Vancouver's Simon Fraser University's School of Kinesiology. While living in Vancouver, Erik gained his pilot's license, started climbing mountains, and took up sky-diving to relax in his

spare time. In 2005, the author worked as an astronaut training consultant for Bigelow Aerospace in Las Vegas and wrote *Tourists in Space*, a training manual for spaceflight participants. He is a Fellow of the British Interplanetary Society and a member of the Aerospace Medical Association. Recently, he was one of the final 30 candidates of the Canadian Space Agency's Astronaut Recruitment Campaign. Erik currently works as manned spaceflight consultant and author. He plans to travel into space with one of the private spaceflight companies. As well as being a triathlete, skydiver, pilot, and author, Erik is an avid scuba-diver and has logged more than 200 dives in more than 20 countries. His favorite movie is the director's cut of *Blade Runner*, his favorite science fiction authors are Allen Steele and Stanislav Lem, and his favorite science fiction series is *Red Dwarf*. *Prepare for Launch* is his fifth book. When not writing, he spends as much time as possible in Kona on the Big Island of Hawaii and at his real home in Sandefjord, Norway. Erik lives with his wife and two cats on the Niagara Escarpment in Canada.

Figures

The colour section appears between pages 156 and 157.

Tables

Panels

Abbreviations

ACES	Advanced Crew Escape Suit
ACT	Advanced Concepts Team
AGSM	Anti-G Straining Manoeuvre
AI	Artificial Intelligence
ALTEA	Anomalous Long Term Effects in Astronauts
AMP	Acoustics Measurement Program
APC	Armored Personnel Carrier
AQF	Astronaut Quarantine Facility
ARC	Ames Research Center
AR&D	Automated Rendezvous and Docking
ATV	Automated Transfer Vehicle
BEES	Bioinspired Engineering of Exploration Systems
BLS	Basic Life Support
CapCom	Capsule Communicator
CBT	Computer Based Training
CEEG	Crew Escape Equipment Group
CEV	Crew Exploration Vehicle
CEVIS	Cycle Ergometer with Vibration Isolation System
CF	Canadian Forces
CHeCS	Crew Health Care Systems
CLL	Central Light Loss
CMO	Crew Medical Officer
CNM	Computer Network Modeling
CNS	Central Nervous System
COL	Crew Options List
CQRM	Crew Qualifications and Responsibility Matrix
CRM	Crew Resource Management
CSA	Canadian Space Agency
CSVS	Canadian Space Vision System
CT	Computer Tomography
CTN	Crew Training Notebook

DAM	Debris Avoidance Manoeuvre
DARPA	Defense Advanced Research Project
DCP	Display and Control Panel
DCS	Decompression Sickness
DMS	Data Management System
DoF	Degrees of Freedom
D-RATS	Desert Research and Technology Study
EAC	European Astronaut Center
ECG	Electrocardiograph
ECLSS	Environmental Control Life Support System
EDL	Entry, Descent and Landing
EDR	European Drawer Rack
EDS	Earth Departure Stage
EEG	Electroencephalograph
EES	Emergency Escape System
EET	Emergency Egress Training
EMU	Extravehicular Mobility Unit
ENT	Ear, Nose, Throat
EPDS	Electrical Power Distribution System
EPM	European Physiology Module
ESA	European Space Agency
EST	Ejection Seat Training
ETC	European Transport Carrier
EVA	Extravehicular Activity
EVCPDS	Extravehicular Charged Particle Directional Spectrometer
FCB	Functional Cargo Block
FCER	Flight Crew Equipment Representative
FCL	Flight Crew Licensing
FCOD	Flight Crew Operations Directorate
FCR	Flight Control Room
FD	Flight Director
FEPA	Flight Equipment Processing Associate
FIT	Final Inspection Team
FOR	Frame of Resolution
FSL	Fluid Science Laboratory
FTS	Flight Termination System
GCR	Galactic Cosmic Radiation
GCTC	Gagarin Cosmonaut Training Center
G-LOC	Gravity-Induced Loss of Consciousness
GLS	Ground Launch Sequencer
GNC	Guidance Navigation and Control
GOR	Gradual Onset Rate
GRC	Glenn Research Center
HAI	High Altitude Indoctrination
HEAT	High-fidelity Environment Training

HMD	Head-Mounted Display
HMP	Haughton Mars Project
HOSC	Huntsville Operations Support Center
HPS	Human Patient Simulator
HRF	Human Research Facility
IBMP	Institute for Biomedical Problems
IGF	Insulin Growth Factor
ILOB	Icarus Lunar Observatory Base
IMS	Inventory Management System
INS	Inertial Navigation System
IOP	Intraocular Pressure
IR	Infrared
ISPR	International Standard Payload Rack
ISRU	In Situ Resource Utilization
ISS	International Space Station
IST	Increment-Specific Training
IVA	Intravehicular Activity
IVCPDS	Intravehicular Charged Particle Directional Spectrometer
JAA	Joint Aviation Authority
JAR	Joint Aviation Requirements
JGTF	Jake Garn Training Facility
JSC	Johnson Space Center
KSC	Kennedy Space Center
LAS	Launch Abort System
LCC	Launch Control Center
LD	Launch Director
LEE	Latching End Effector
LEO	Low Earth Orbit
LH2	Liquid Hydrogen
LLOX	Lunar Liquid Oxygen
LMA	Laryngeal Mask Airway
LMESSC	Lightweight Multi-Purpose Experiment Support Structure Carrier
LOV	Loss of Vision
LOX	Liquid Oxygen
LPS	Launch Processing System
LRC	Langley Research Center
LSAH	Longitudinal Study of Astronaut's Health
MAG	Maximum Absorbency Garment
MBS	Mobile Base System
MCAT	Medical College Admission Test
MCC	Mission Control Center
MCP	Mechanical Counter-Pressure
MDD	Multispatial Disorientation Device
MER	Mission Evaluation Room
MIT	Massachusetts Institute of Technology

MLC	Multimedia Learning Center
MMPI	Minnesota Multiphasic Personality Inventory
MMT	Mission Management Team
MOI	Mars Orbit Insertion
MOL	Manned Orbiting Laboratory
MRI	Magnetic Resonance Imaging
MRO	Mission Robotics Operator
MSFC	Marshall Space Flight Center
MSFT	Multi-Stage Fitness Test
MSS	Mobile Servicing System
MSSOTS	Mobile Servicing System Operations Training Simulator
NBL	Neutral Buoyancy Laboratory
NEEMO	NASA Extreme Environment Mission Operations Project
NSBRI	National Space Biomedical Research Institute
NTD	NASA Test Director
OBS	Operational Bioinstrumentation System
OETF	Operations Engineering Training Facility
OFK	Official Flight Kit
OID	Operational Intercommunications System
PBAN	Polybutadiene Acrylonitrite
PDGF	Power Data Grapple Fixture
PLL	Peripheral Light Loss
PLSS	Portable Life Support System
PPC	Private Psychological Conference
PPK	Personal Preference Kit
RDS	Russian Docking System
RHC	Rotational Hand Controller
ROR	Rapid Onset Rate
ROV	Remotely Operated Vehicle
RRT	Rapid Response Team
RSO	Range Safety Officer
RSS	Rotating Service Structure
RWS	Robotic Work Station
SAR	Synthetic Aperture Radar
SARJ	Solar Alpha Rotary Joint
SCA	Simulation Control Area
SCTF	Sonny Carter Training Facility
SETI	Search for Extraterrestrial Intelligence
SFU	Simon Fraser University
SM	Service Module
SOLO	Sodium Loading in Microgravity
SOMD	Space Operations Mission Directorate
SPE	Solar Particle Event
SRB	Solid Rocket Booster
SRC	Short-Radius Centrifuge

SSC	Stennis Space Center
SSMTF	Space Station Mockup and Training Facility
SSPF	Space Station Processing Facility
SST	Single System Trainer
SSTF	Space Station Training Facility
STFO	Spaceflight Training and Facility Operations
STL	Station Training Lead
SVMF	Space Vehicle Mock-up Facility
SVP	Structural Verification Plan
SWG	Structures Working Group
TBA	Trundle Bearing Assembly
TCDT	Terminal Countdown Demonstration Test
TCM	Trajectory Correction Manoeuvre
TCS	Thermal Control System
TEI	Trans-Earth Insertion
THC	Translational Hand Controller
TIC	Thermal Imaging Camera
TLI	Trans-Lunar Insertion
TORU	Tele-Operator Control System
TPS	Thermal Protection System
TVIS	Treadmill with Vibration and Isolation System
UPA	Urine Processing Assembly
USA	United Space Alliance
USAF	United States Air Force
VAB	Vehicle Assembly Building
VEG	Virtual Environment Generator
VR	Virtual Reality
VSE	Vision for Space Exploration
WRS	Water Recovery System
ZSF	Zvezda Space Facility

Section I

Astronaut Selection

When the first Mercury astronauts dazzled the world with their spectacular achievements, they became instant heroes. To the admiring public, the job of astronaut appeared to be a glamorous one, combining the skills of a fighter pilot and the tenacity of an explorer with the fame normally reserved for movie stars. This perception was reinforced by the fact many of the early flights routinely concluded with visits to the Oval Office and ticker-tape parades. However, despite astronauts' spending so much time in the public eye, few people ever stopped to wonder how the role of the astronaut came to be defined and how the application and selection requirements were developed. Even today, after several manned spaceflight programs, and despite the job of astronaut being perhaps the most well known of all federal jobs, few people understand how astronauts are selected and trained.

Section I first examines the history of astronaut recruitment and the selection process from the perspectives of NASA, the European Space Agency (ESA), and the Canadian Space Agency (CSA). The section then examines the current application procedures, before focusing on the process of selecting the astronauts who will eventually embark upon missions to the Moon and Mars.

1

A brief history of astronaut selection

NASA

NASA began its search for astronaut candidates shortly after the space agency was created in July, 1958. In October, 1958, the father of American manned spaceflight, Robert R. Gilruth, received approval for Project Mercury, a program requiring the creation of an astronaut corps. Selection criteria for the first American astronauts were quite simple, since President Dwight D. Eisenhower directed that the astronauts be selected from the military's pool of test pilots. Due to the inherent dangers of spaceflight (Figure 1.1 and Table 1.1), and the potential security implications of the program, the decision to only select military pilots was a logical one, as concluded by space historian, Margaret Weitekamp:

> "From the military test flying experience, the jet pilots also mastered valuable skills that NASA wanted its astronauts to possess. Test pilots were accustomed to flying high-performance aircraft, detecting a problem, diagnosing the cause, and communicating that analysis to the engineers and mechanics clearly. In addition, they were used to military discipline, rank and order. They would be able to take orders. Selecting military jet test pilots as their potential astronauts allowed NASA to choose from a cadre of highly motivated, technically skilled, and extremely disciplined pilots."[1]

Table 1.1. Hazards of manned spaceflight.

Equipment malfunction	High gravitational forces
Turbulence and impact forces	Hazards of emergency egress
Impact forces during launch and re-entry	Effects of radiation exposure
Physiological effects of microgravity	Effects of temperature fluctuations
Psychological effects of isolation and confinement	Rapid and/or explosive decompression of spacecraft
Fire and explosions	Crash due to pilot error

Figure 1.1. Due to the hazards of spaceflight and security implications of the Mercury and Gemini Programs, the first astronauts selected were military pilots. In this photo, Ed White, an Air Force Lieutenant Colonel recruited for the Gemini Program, performs America's first spacewalk. Image courtesy: NASA.

Selecting the Mercury astronauts

Part of NASA's first astronaut description:

> "Although the entire satellite operation will be possible, in the early phases, without the presence of man, the astronaut will play an important role during the flight. He will contribute by monitoring the cabin environment and making necessary adjustments. He will have continuous displays of his position and

> attitude and other instrument readings, and will have the capability of operating the reaction controls, and of initiating the descent from orbit. He will contribute to the operation of the communications system. In addition, the astronaut will make research observations that cannot be made by instruments; these include physiological, astronomical and meteorological observations."

Based on NASA's job description, many experienced pilots believed NASA was recruiting passengers, not pilots. Also, since rockets were considered to be an especially risky mode of travel, some pilots dubbed the astronauts "ham in a can". Nevertheless, more than 500 military pilots responded to NASA's first request for astronaut candidates. From a total of 508 service records screened in January, 1959, NASA found 110 that met the minimum standards established for Mercury:

- Age – less than 40
- Height – less than 1.80 m
- Excellent physical condition
- Bachelor's degree or equivalent
- Graduate of test pilot school
- 1,500 hr total flying time
- Qualified jet pilot.

At the time of Project Mercury, there were no standardized tests for selecting aspiring astronauts. Although tests were developed during the first selection and subsequent recruitment campaigns, Project Mercury was perhaps most notable for the role played by military hierarchies. In fact, some prospective candidates were eliminated as the military services vied for a balance between the different branches of the services represented. In some cases, admirals and generals even sponsored their favorite candidates and some top-ranking leaders took acceptance or rejection of their candidate personally.

Each of the 110 candidates went to the Lovelace Clinic in Albuquerque, New Mexico, for medical testing. The medical assessment included encephalography, cardiography, and X-ray examinations in addition to ophthalmology and otolaryngology evaluation. Physiological examinations included bicycle ergometer tests, calculation of the specific gravity of the body, and even total-body radiation count. At the end of the series of tests, the Lovelace physicians had probably performed the most complete medical assessments in history.

Next for the Mercury candidates came an elaborate series of environmental studies, physical endurance tests, and psychiatric studies, conducted at the Aeromedical Laboratory of the Wright Air Development Center, Dayton, Ohio. During March, 1959, candidates were subjected to an array of evaluations, including pressure suit tests, vibration tests, noise tests, not to mention a barrage of psychological tests designed to assess personality and motivation. Occasionally, the candidates would play games with their evaluators – a tactic that may have resulted in Charles "Pete" Conrad (Figure 1.2) being deselected for Project Mercury (he was eventually selected as an astronaut in 1962 for Project Gemini). For example, one of the psychological evaluations involved a psychologist showing a candidate a blank

Figure 1.2. Pete Conrad was initially considered for the Mercury Program, but was eventually selected for Project Gemini. Image courtesy: NASA.

white card and asking him what he saw. Conrad, when presented with the test, simply answered that the card was upside down! On another occasion, when asked to deliver a stool sample to the on-site lab, Conrad placed it in a gift box and tied a red ribbon around it! Eventually, he decided he'd had enough. After dropping his full enema bag on the desk of the lab's commanding officer, Conrad walked out. His

Project Mercury application was subsequently denied, with the notation "not suitable for long-duration flight".

Following the tests in Dayton, NASA whittled the group down to just 18 candidates, and, after evaluation of technical qualifications against the technical requirements of the program, seven men were chosen. They were presented to the public as the first Americans to fly into space on April 9th, 1959, and became heroes almost immediately:

- Lt. Col. John H. Glenn Jr. (Marine Corps)
- Lt. Cdr. Walter M. Schirra, Jr. (Navy)
- Lt. Cdr. Alan B. Shepard, Jr. (Navy)
- Lt. M. Scott Carpenter (Navy)
- Capt. L. Gordon Cooper (Air Force)
- Capt. Virgil I. "Gus" Grissom (Air Force)
- Capt. Donald "Deke" Slayton (Air Force).

The fame of the Mercury 7 quickly grew beyond all proportion to their activities, which wasn't surprising given the high profile of the program. The fundamental purpose of Project Mercury was to determine whether humans could survive the demands of launch and orbit in the inhospitable environment of space. From this perspective, the astronauts weren't comparable to traditional explorers like Shackleton, Byrd, or Hillary. In fact, in the eyes of the public and the media, the astronauts left conventional explorers in the shade.*

Female astronauts

In 1960, NASA began a program to determine whether women could qualify as astronauts. Twenty-five female pilots were invited by NASA to undergo the same physical and psychological evaluations as those endured by the male astronauts who had qualified for the Mercury program. Of those invited, 13 women passed the tests and were enrolled in an "unofficial" astronaut training program. While the 13 were subjected to many of the same tests, evaluations, and training as their male counterparts, these women were never officially declared astronaut candidates, nor were they scheduled for space flights. NASA management reportedly believed that damaging negative publicity would be generated if a woman were injured or killed during a spaceflight. It was also believed the participation of female astronauts in the early space program would divert scarce resources and attention away from male astronauts. Ultimately, the initial NASA training of female astronauts ceased in 1963, and the program quickly faded into history. Ironically, the same year, the Soviet Union launched the first woman, Valentina V. Tereshkova, into space.

---

* The mythological dimension of the job of an astronaut was later captured in Tom Wolfe's epic novel, *The Right Stuff*, a book that accurately described how the creation of the astronaut corps helped galvanize support for the manned spaceflight program they represented.

From Gemini to Shuttle

The single-astronaut Project Mercury paved the way for the more ambitious two-astronaut Project Gemini, requiring NASA to issue a second call for astronaut candidates. While the qualifications for the Gemini astronaut candidates remained mostly the same as those for the Mercury 7, the criteria were broadened slightly. For example, the maximum height requirement was raised from 5 foot, 11 inches to an even 6 foot, minimum flight time was reduced from 1,500 to 1,000 hr, and maximum age was reduced to not older than 35. For the first time, civilian and female pilots were accepted as astronaut applicants but, while this and future astronaut drafts were open to women, no women were selected as astronaut candidates until 1978. From the second astronaut draft, nine were selected.

USAF X-20 "Dyna-Soar"

While the NASA manned spaceflight program was still in its infancy, the United States Air Force (USAF) had begun its own research program developing a manned multi-use orbiting spaceplane. The X-20 was nicknamed the "Dyna-Soar", short for "dynamic soaring". Dynamic soaring is a principle whereby an object can achieve high altitudes using a combination of rocket propulsion and aerodynamic lifting ability. Although the project was canceled before a single test flight was conducted, the USAF assigned six Pilot Astronauts to the project.

NASA Group 3

While Project Gemini was under development, NASA prepared for the more ambitious three-astronaut Apollo program. Apollo was designed to carry astronauts to the Moon and back, which required even more astronauts, so a third astronaut selection campaign was issued. Requirements for the budding Apollo astronaut candidates were identical to those selected in NASA Group 2. From a total of about 400 astronaut candidates, 14 were selected.

NASA Group 4

As the Apollo program evolved, NASA faced an increasing tide of criticism that the astronaut corps was dominated by military pilots. Responding to this disapproval, NASA issued a call for astronaut candidates who would serve as "Scientist Astronauts". While basic selection requirements were similar to those of previous selection campaigns, the minimum flight time criterion was waived. Instead, scientist astronauts were required to possess a doctorate degree in the natural sciences, medicine, engineering, or equivalent experience. About 1,500 Scientist Astronaut candidates applied for NASA's fourth astronaut campaign, of which several were rejected immediately because they wore glasses. Ultimately, only six Scientist Astronauts were selected.

USAF Manned Orbiting Laboratory

Just as they had attempted to develop the first spaceplane, the USAF also attempted to develop the first US space station, known as the Manned Orbiting Laboratory (MOL), but, like the X-20, the MOL was destined never to fly.

The program was officially acknowledged by Secretary of Defense, Robert McNamara, on December 10th, 1963 – a date which, ironically, was the same day McNamara axed the X-20 program. The USAF planned for the MOL to be launched and operational by 1971. Envisioned as a laboratory module, the MOL would provide a crew of two astronauts with a shirt-sleeve working environment. Astronauts were to be ferried to and from the MOL by a modified Gemini incorporating a hatch in the aft heat shield that would be mated to a crew transfer tunnel on the MOL. Since the MOL involved manned space flight, the program required Pilot Astronauts and so more astronauts were selected. The requirements for MOL Pilot Astronauts were quite similar to those of NASA's Pilot Astronaut, although, once selected, the MOL astronauts were sent back to the Aerospace Research Pilot School for training suited to the MOL program, rather than regular astronaut training.

NASA Group 5

As the Apollo program progressed, NASA required more astronaut pilots and a fifth astronaut selection campaign was conducted, from which 19 candidates were selected.

USAF MOL Groups 2 and 3

As the USAF prepared for its first planned series of MOL research and development test flights, a second group of MOL Pilot Astronauts was drafted. With two groups of MOL Pilot Astronauts already in training, the first test flight of the MOL program was successfully conducted. On November 3rd, 1966, a Titan III-C was launched from Cape Canaveral carrying an unmanned Gemini capsule. The Titan III-C carried the capsule to space before propelling it on a high-speed ballistic re-entry trajectory. The success of the first MOL mission, coupled with the momentum of the program, prompted the USAF to draft a third group of Pilot Astronauts.

Scientist Astronaut

Since NASA had been disappointed with the quality of Scientist Astronaut candidates in the fourth astronaut draft, another selection campaign was instigated

Figure 1.3. Bob Crippen examines a spacesuit designed for the Manned Orbiting Laboratory program. Crippen was selected as a MOL astronaut before the program was cancelled. He later flew on the first Space Shuttle mission for NASA in April, 1981. Image courtesy: NASA.

to select more astronauts with scientific expertise. From this campaign, 11 candidates were selected. Meanwhile, the MOL program was abruptly canceled in 1969, since the Pentagon judged the program would not perform relative to its high cost, and would pose a threat to NASA scientific programs under development at the same time. Seven of the MOL Pilot Astronauts were transferred to NASA, some of whom, such as Bob Crippin (Figure 1.3), would go on to make significant contributions in the development and evolution of the Space Shuttle program.

Space Shuttle astronauts

The dawn of the Space Shuttle (Figure 1.4) program required a fresh crop of astronaut candidates. While NASA had maintained a number of astronaut holdovers from previous programs, an ambitious initial forecast of up to 60 Shuttle missions per year* required more astronauts than were available.

NASA grouped Shuttle astronauts into three categories. These included Pilot Astronauts, Mission Specialists, and Payload Specialists. Pilot Astronauts were

* The highest number of shuttle missions flown in one year was nine, in 1985.

Figure 1.4. The dawn of the Space Shuttle program required a fresh crop of astronauts, beginning with the NASA Group 8 in 1978. Image courtesy: NASA.

further divided into Space Shuttle Commanders and Space Shuttle Pilots. The selection criteria established for the Space Shuttle astronauts have changed little since NASA Group 8 in 1978, which selected the first astronauts to fly on the Shuttle.

The selection criteria for Pilot Astronauts require applicants to possess a bachelor's degree from an accredited institution in engineering, biological science, physical science, or mathematics. An advanced degree is also desirable, as are at least 1,000 hr of pilot-in-command flight time, and flight test experience. Pilot Astronauts are also required to pass a NASA Class I space physical, including a minimum 20/50 uncorrected vision, correctable to 20/20, and a maximum blood pressure of 140/90 mmHg in the sitting position. Height requirements are a minimum of 5 foot, 4 inches and a maximum of 6 foot, 4 inches.

Mission Specialists must possess a bachelor's degree from an accredited institution in engineering, biological science, physical science, or mathematics. This degree must be followed by at least three years of related professional experience, although advanced degrees may be substituted for professional experience. Mission Specialist candidates must pass a NASA Class II space physical, including a minimum 20/150 uncorrected vision, correctable to 20/20, and a maximum blood pressure of 140/90 mmHg in a sitting position. As per federal regulation, NASA is not allowed to specify an age range for astronaut candidates.

Payload Specialists are not maintained as NASA astronauts but are employed by NASA for individual missions. Payload Specialists possess knowledge and expertise unique to a particular payload and train along with an assigned crew for each mission. For that reason, specific qualifications for Payload Specialists vary. Astronauts flown as Payload Specialists (some would argue as passengers) have included US Senator Jake Garn, US Representative Bill Nelson, and US Senator John Glenn.

CANADIAN SPACE AGENCY

In 1983, following an offer by NASA to fly Canadian astronauts as Payload Specialists aboard the Space Shuttle, a screening committee from the National Research Council of Canada (NRC) conducted Canada's first astronaut selection campaign, resulting in the selection of six astronauts. Canada's second group of astronauts was selected in June, 1992, following a five-month selection process conducted by the Canadian Space Agency (CSA). Four astronauts were selected from this campaign. In May, 2008, the CSA launched its third selection campaign to recruit two new members to the Canadian Astronaut Corps. The agency received 5,351 applications for the year-long selection process – the most rigorous and demanding selection yet employed by a national space agency. A year later, on May 13th, 2009, the CSA announced the successful candidates who commenced training with NASA's new astronaut class in August 2009.

EUROPEAN ASTRONAUT SELECTION

In 1978, Europe created its own human spaceflight history when Sigmund Jähn from Germany (former GDR) became the first European in space after being launched on the Russian Soyuz 31 spacecraft to the Salyut 6 space station. The European Space Agency (ESA) conducted its first astronaut selection in 1977–1978. The campaign originated from a 1973 agreement ESA had with NASA to supply the first Spacelab reusable science laboratory (carried in the Shuttle's cargo bay) in exchange for flight opportunities for European astronauts. From the 1977–1978 selection campaign, ESA chose three astronauts, one of whom was Ulf Merbold, who had the honor of being the first ESA astronaut to undertake a mission on the Space Shuttle (STS-9) on the 10-day Spacelab-1 mission.

During the 1980s, while ESA astronauts were undertaking Shuttle missions, other Member States of ESA conducted their own astronaut recruitment. Many of those astronauts recruited by these national programs flew on Russian Soyuz missions to the Mir Space Station or on American Space Shuttle missions as Payload Specialists. In 1990, ESA recognized the need to increase the size of its Astronaut Corps and the European Astronaut Centre (EAC) was founded. This was in response to projects and studies that would eventually be aligned with Europe's contribution to the International Space Station (ISS), including the *Columbus* laboratory, which was launched and attached to the ISS in February, 2008.

In 1991–1992, a new astronaut selection was conducted. First, there was a national selection under the purview of each ESA Member State that had the opportunity to present up to five candidates to ESA for final selection. Collectively, the Member States received more than 22,000 applications, from which nearly 5,500 met initial selection criteria. Following a national psychological, medical, and professional screening process, ESA identified 59 candidates for the second step, the final ESA selection. From the final group, ESA recruited six astronauts. Four of them started basic training at EAC and were later trained for Russian Soyuz flights to the Russian Mir space station. The other two were sent to NASA to start Shuttle Mission Specialist training, being incorporated into the NASA astronaut corps.

In 1998, the year witnessing the launch of the first two elements of the ISS, ESA decided to integrate most of the national astronauts into a single European Astronaut Corps. Ten years later, the number of astronauts in the European Astronaut Corps had dwindled to just eight due to astronauts retiring and assuming other duties within ESA. With manned missions to the Moon planned, ESA decided it was time for another astronaut selection campaign. ESA's 2008 astronaut selection campaign was the first time a selection had been performed by ESA all over Europe on a web-based application procedure without any national pre-selection. More than 8,000 applications were received, from which six astronauts were selected in May, 2009. With the prospect of future human exploration missions to the Moon and Mars, ESA has already indicated its plans for another astronaut selection campaign in 2014.

While the first astronaut groups were exclusively male military test pilots, today's astronaut groups are an integrated group of pilots, scientists, and mission specialists. However, all astronauts, whether they were chosen 50 years ago or were selected in the latest round of recruitment campaigns, were chosen because they were highly capable, highly skilled, and highly trained. They belong to a unique and elite group who rightfully identify themselves with the pride anyone would feel in belonging to such an esteemed group and, as we shall see in the next chapter, the challenges of becoming a member are just as difficult today as they were in the days of John Glenn and Pete Conrad.

REFERENCE

1. Weitekamp, M.A. The Right Stuff, The Wrong Sex: The Science, Culture, and Politics of the Lovelace Woman in the Space Program, 1959–1963, Ph.D. Diss., Cornell University, p. 98 (2001).

2

Applying to become an astronaut

> "The Canadian Space Agency (CSA) is looking for individuals who want to be part of the next generation of space explorers. Two applicants will be selected to join the CSA Astronaut corps. A pool of qualified candidates will also be created for future needs.
>
> The CSA is seeking outstanding scientists, engineers and/or medical doctors with a wide variety of backgrounds. Creativity, diversity, teamwork, and a probing mind are qualities required to join the CSA's Astronaut Corps. To withstand the physical demands of training and space flight, candidates must also demonstrate a high level of fitness and a clean bill of health."
>
> Advertisement posted by the CSA, June 26th, 2008

It's the world's most unusual want ad. Say the word "astronaut", and people immediately think of Neil Armstrong walking on the Moon or astronauts being launched atop the Space Shuttle. It's a job that has always conjured up visions of heroes and heroic feats. But being an astronaut is not all glamour and glory. And, more often than not, being an astronaut is not about spending that much time in space. In fact, the most time spent in space by one astronaut (Sergei Krikalev) is just 803 days, which works out to be just short of 2.5 years. When you consider many people spend 30 or more years in their professional career, 2.5 years doesn't sound that much. What do astronauts do with the rest of their time?

Most astronauts are civil servants. As civil servants, they have to attend meetings, go to training sessions, and write reports, just like any other office worker. They do, however, possess several specialized skills unique to their trade. And they enjoy, albeit rarely, opportunities to travel and work in space! From that perspective, astronauts could be considered as regular, ordinary government employees who get to travel extensively, both around the world and in space. However, in reality, there is nothing regular about this job, and no occupation could be less ordinary.

First, consider the risk. Of the 500 people* who have been launched into space as

---

* The 500th person in space was NASA mission specialist, Chris Cassidy, a Navy SEAL-turned-astronaut, who launched onboard the Space Shuttle *Endeavour*'s STS-127 on July 15th, 2009.

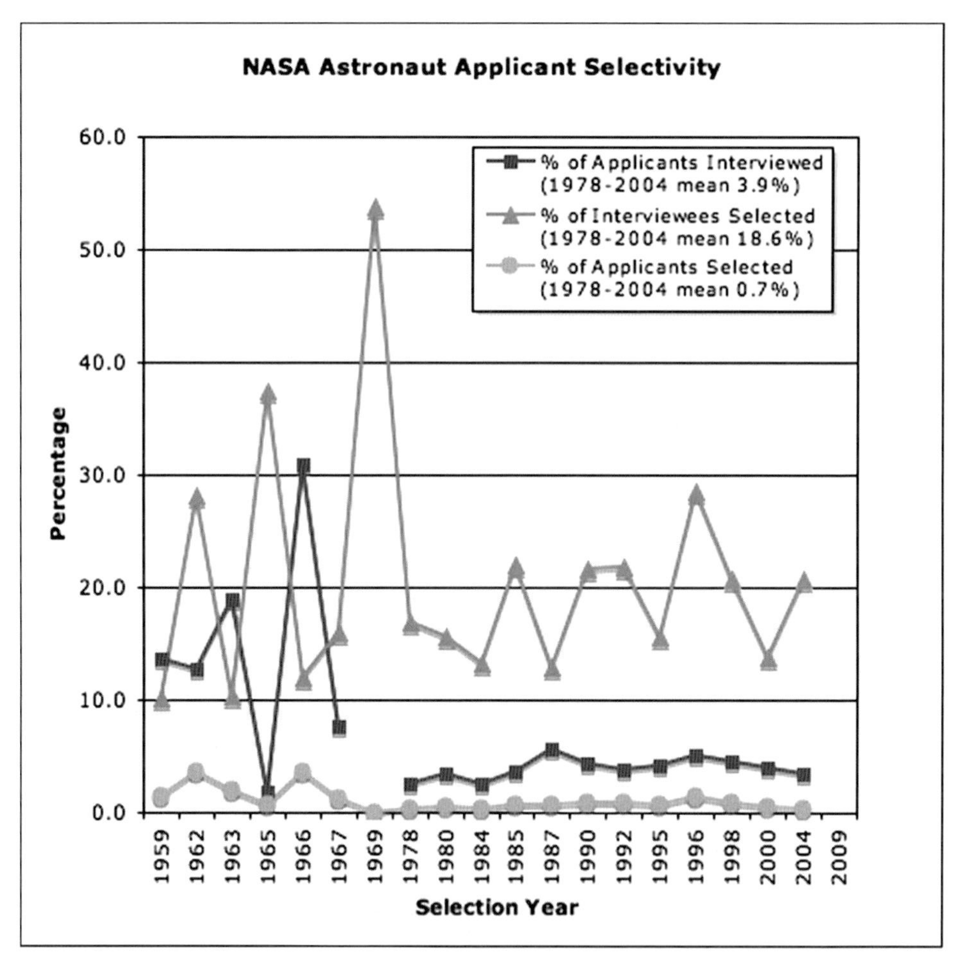

Figure 2.1. NASA astronaut applicant selectivity. Image courtesy: NASA.

of July, 2009, 18 have died during the mission. This mortality rate of 3.74% makes being an astronaut one of the most dangerous professions anyone could consider.* By comparison, American combat troops suffered a 0.39% mortality rate in Iraq between 2003 and 2006 and 2.18% in Vietnam between 1966 and 1972. Apart from the horrendous risks, the job comes with crazy work hours, grinding travel (albeit, some of it in space!) and decent, though not exceptional, pay. Added to this less-than-desirable work description are the daunting odds of selection, which inevitably fuels the question: Why?

Perhaps the most dominant factor driving wannabe astronauts to submit to a year-long recruitment and selection process is simply a love for space, followed a

* The risk is equivalent to 400 Boeing 747s crashing every day, or 197,000 passengers being killed every day!

close second by the desire to contribute to a great endeavor. A third motivating factor is being able to work with exceptional people who share the same vision and to play a part in continuing to push outward the boundaries of human space travel. Given these motivating factors, the risks and the long hours are considered minor inconveniences. This is also why, despite the occupational challenges, and the odds of being selected notwithstanding (Figure 2.1), NASA, the Canadian Space Agency (CSA), and the European Space Agency (ESA) have never had a problem filling their astronaut ranks.

BASIC QUALIFYING CRITERIA

Those serious about submitting an application have usually spent at least a decade or more acquiring the qualifications necessary to be considered. However, a stellar education background isn't sufficient to catch the eye of the selection committee. The multitude of tasks astronauts must perform as part of their job demands multitalented people.

Education

Although the minimum educational requirements for astronaut application include a bachelor's degree (Panel 2.1), in reality, the minimum requirement for a competitive application is a Ph.D. (some candidates have a Ph.D. *and* a medical degree!), unless you happen to be a pilot. In fact, just having a Ph.D. isn't sufficient to guarantee a candidate will progress to the preliminary interview phase. To have any chance of being considered a serious candidate, a Ph.D. needs to be coupled with an outstanding professional background in research. If the applicant has aspirations to be one of the 15–20 candidates selected for the final panel interview, even an outstanding professional background is inadequate. For the selectors to take note, the applicant has to indicate in their resume they have published research papers, perhaps authored a few books, presented papers at conferences, pioneered new research, and maybe even filed the odd patent or two.

Health and fitness

Equally essential is excellent physical condition. Astronauts must undergo intensive periods of training and may participate in spaceflights lasting 6 months or more, during which their body will be subject to a great deal of stress such as bone demineralization and muscle atrophy. Being able to deal with this stress obviously requires a healthy body and a sound mind, so applicants need to have above average health. While some medical standards were relaxed during the Shuttle era, due to the short durations of typical Shuttle missions, with the dawn of Expedition-Class missions to the ISS, medical standards will be more stringent than ever. These

Panel 2.1. CSA general selection requirements

- The applicant must be a Canadian citizen or a resident of Canada.
- Because astronauts are required to perform a broad range of scientific and technical work, prospective candidates must hold a bachelor's degree recognized in Canada, in one of the following areas:
  - Engineering or Applied Sciences.
  - Science (e.g. Physics, Chemistry, Biology, Geology, Mathematics, Computer Science, etc.).
  - The bachelor's degree must be followed by at least two years of related professional experience.

    OR:

    A bachelor's degree along with a master's degree or a doctoral degree recognized in Canada, in one of the following areas:
  - Engineering or Applied Sciences.
  - Science (e.g. Physics, Chemistry, Biology, Geology, Mathematics, Computer Science, etc.).

    OR:
  - A license to practice medicine in a province or a territory of Canada.

Medical requirements

To be selected, applicants must meet stringent medical criteria. Applicants will be required to undergo Canadian Space Agency medical physical exams, which include the following specific requirements:

- Standing height must be between 149.5 and 190.5 cm.
- Visual acuity must be 20/20 (6/6) or better in each eye, with or without correction.
- Maximum limits for cycloplegic refractive error and astigmatism correction also apply.
- The refractive corrective surgical procedures PRK or LASIK are allowed. For those candidates under final consideration, an operative report on the surgical procedure will be requested. The CSA does not recommend that potential candidates undergo laser refractive surgery for the sole purpose of applying for employment as an astronaut.
- Blood pressure must not exceed 140/90 mmHg, measured in a sitting position.
- Meet the following pure tone audiometry hearing thresholds:
  Frequency (Hertz): 500 1000 2000 3000 4000
  Either ear (decibels): 30 25 25 35 50

Conditions of employment

- The position will require frequent travel and relocation.
- The position is subject to pre-employment security clearance.
- Candidates must undergo pre-employment medical examinations.

standards simply reflect the fact that statistical risk for a medical event increases with mission duration. For example, an episode of renal colic in a mission specialist may have been acceptable for a short-duration Space Shuttle mission, but is disqualifying for an Exploration-Class mission. This is because pre-flight ultrasound screening can rule out significant retained calculi and the probability of developing a calculus during a 10–14-day Shuttle mission is very low. While most people assume they are healthy, it is worth highlighting some statistics from the CSA's 1992 astronaut recruitment campaign (Table 2.1), which sent out 337 questionnaires to the most qualified applicants.

Table 2.1. Medical disqualification during 1992 CSA astronaut recruitment.*

Reason for disqualification	*Number*	*Percent disqualified*
Vision	105	31
Migraine history	12	3.6
Thyroid disorders	5	1.5
Ears/hearing	4	1.2
Lungs/asthma	3	0.8
Misc. (1 each), inc. Hodgkin's, MS, Crohn's, epilepsy, obesity, vertigo, others	16	4.7
Totals	145	43

* Based on 337 medical questionnaires received during the CSA's 1992 Recruitment Campaign.

Miscellaneous experience

In the resume I submitted as part of my astronaut application, I entitled this section *avocational experience*. Perhaps one of the most important skills space agencies look for after education is flying experience. At the very minimum, they expect applicants to possess at least a private pilot's license (eight of the 16 final candidates in the CSA selection were professional pilots). If applicants happen to have a multi-engine and/or instrument rating in addition to a commercial rating, so much the better. Previous experience with aircraft operations is a bonus, particularly if it involved tasks such as being a test pilot or flight engineer.

Next, agencies look for skills relevant to the job of being an astronaut. Parachuting experience is an asset, since, in the event of some abort contingencies, this is a skill astronauts may have to use. Another skill used extensively by astronauts is scuba-diving when practicing in the Neutral Buoyancy Laboratory (NBL). In fact, for those few astronauts who do not have scuba-diving certification, this is one of the first things they learn during basic training.

DESIRABLE QUALITIES

In addition to the veritable shopping list of qualifications a potential astronaut candidate must possess, space agencies also look for certain qualities they consider make up an ideal spacefarer.

Patience

One virtue required by all astronauts is patience. Having spent more than a decade accumulating the necessary qualifications for the application process, most applicants have already demonstrated this quality! In addition to embarking upon a lengthy probationary training schedule, astronauts must also be prepared to spend several years working various technical assignments before being selected for a mission, which means more training lasting 18 months or more. By the time launch day finally rolls around, some astronauts have waited 10 years or more since being selected (Robert Thirsk (Figure 2.2) waited nearly 13 years for his first flight!). Even those lucky to fly three or four times in a career have routinely waited 5 or 6 years between flights. In the post-Shuttle era, possessing a reservoir of patience will be even more important due to even fewer flight opportunities being available. While the Shuttle was flying, new astronauts could expect to make two to four flights in a career that normally stretched 10–12 years. In some years, as many as 50 astronauts flew in space, but with the retirement of the Shuttle in 2010 and the advent of ISS Expedition increments, that number will be more like a dozen. For the new generation of astronauts, a typical career may offer only one or two flight opportunities. It might sound like a lot of work for little reward, but, judging by the number of applications, the lack of flight opportunities had little impact upon the recruitment campaigns conducted by NASA, the CSA, and the ESA in 2008–2009.

Versatility

> "A human being should be able to change a diaper, plan an invasion, butcher a hog, conn a ship, design a building, write a sonnet, balance accounts, build a wall, set a bone, comfort the dying, take orders, give orders, cooperate, act alone, solve equations, analyze a new problem, pitch manure, program a computer, cook a tasty meal, fight efficiently, die gallantly. Specialization is for insects."
>
> Robert A. Heinlein, *Time Enough for Love*

While an astronaut doesn't need to be proficient in all the tasks itemized by Heinlein, versatility is a key quality demanded by space agencies recruiting astronauts. NASA, the CSA, and the ESA look for versatile individuals who have high technical competence and the ability to work in a team. The selection team also looks for a hybrid: a researcher who can handle the science and operate manipulators such as

Figure 2.2. The wait for a flight can be a long one. Canadian Space Agency astronaut, Bob Thirsk, shown here during astronaut training, was selected in December, 1983 but had to wait until June, 1996 for his first flight. Image courtesy: CSA.

the Canadarm2 and who also has the physical ability to endure grueling spacewalks. Given these requirements, it is tough for applicants to meet the grade and show they possess all those qualities. But the myriad qualifications are necessary, since training an astronaut is a considerable investment for any agency and the support needed both before and during a space mission is costly; for example, the Russians will be charging NASA $51 million a ticket*(!) onboard their Soyuz following the Shuttle's retirement. Furthermore, it takes years to plan and organize a space mission and

* In comparison, Space Adventures charges rich businessmen $30 million for the same flight.

hundreds of people are involved in preparing the astronauts and the spacecraft; since astronauts are pivotal to the success of a mission and flight opportunities are so limited, space agencies obviously want to ensure those selected will make the best possible use of the precious time they will spend in space.

High achiever

A versatile, patient, certified pilot, sky-diver, scuba-diver with a Ph.D. who also possesses the requisite myriad qualifications described earlier is off to a good start, but even this list of credentials won't be sufficient to even guarantee an interview. Those selected as astronauts also need to be high achievers. You might be thinking that collecting all those qualifications would surely put a person in that category, but the agencies are looking for something extra; they're looking for some noteworthy accomplishment outside their professional field of expertise. For example, during the CSA campaign, there was one applicant who had climbed Mount Everest in addition to being a physician and there was another who had raced in the Olympics as well as being a fighter pilot.

Psychological disposition

Another quality the agencies pay particular attention to is psychological disposition. The new class of astronauts will spend several months in a confined space with five other crewmembers. Needless to say, the psychological qualities required will include an ability to get on well with other crewmembers and an affinity for team work and adaptability. To have a chance of making it to the final panel interview therefore, aspiring astronauts need to demonstrate an equable temperament and a high level of self-control. From an operational perspective, agencies seek those with an ability to adapt quickly to changing situations and those who exercise mature judgment, since these are the qualities that will be helpful in performing tasks and optimizing on-orbit routines and procedures. Agencies also look for someone who manages his/her adrenaline well and in a flight situation where they are put into a dangerous situation, does the same thing. For some applicants, such as fighter pilots and emergency physicians, psychological aptitude is already a proven quality; it was no coincidence the two candidates chosen by the CSA in 2009 happened to be a pilot and a physician!

Public relations

Astronauts spend only a fraction of their time in space and when they're not training for their next mission, they are often doing public relations for the space agency. The role of an astronaut as a public relations officer is a natural one, since the media and the public are naturally curious about their life and missions. This means astronauts must enjoy meeting the public and the press, and be able to communicate the importance of their tasks in space. Since astronauts are often the public face of the agency they represent, personality inevitably is a significant selection criterion when it comes down to the final few candidates. Not only must astronauts serve as ambassadors for their agency and country internationally, they must also take complex issues of science, technology, and exploration and explain them in a way that excites young kids at school.

In addition to being able to communicate in their mother tongue, astronauts are also expected to have language ability in at least one other language. Language skills are necessary, given the international collaboration that exists between space agencies today. Fluency in English is mandatory and a good knowledge of Russian is an asset, as it facilitates training at the cosmonaut center in Russia. Interest in, and knowledge of, American, Russian, and Japanese culture is also useful, as it facilitates good relations with international partners on the ISS.

THE APPLICATION PROCESS

Once a candidate has acquired the basic qualifying criteria and considers they have the desirable qualities, they can begin the arduous application process. Even for those who have had their sights set on becoming an astronaut for decades, the application process can appear daunting, especially if you happen to be a non-US citizen (the odds of being selected as a CSA or ESA astronaut are by more than 10 orders of magnitude worse). Having spent the best part of two decades or more accumulating the experience and qualifications just to be able to submit a competitive application, applicants must then undergo a series of preliminary interviews, physical and psychological tests in addition to all sorts of other evaluations. Even for these supremely qualified applicants, the chances of actually becoming an astronaut are a statistical long shot, but not as long as someone who doesn't apply; if you don't submit an application, you have a 100% chance of not making it. That's an inviolable statistic!

Once the application has been submitted, it's a case of waiting for the call to the first stage of selection. While space agencies look for the best people possible, their methods of selection differ. NASA, for example, conducts a panel interview together with routine fitness, medical, and coordination tests, which is an approach similar to the selection performed by the ESA. In contrast, the CSA, which has by far the most rigorous selection of any agency in the world,* requires applicants to submit to myriad tests, many of which wouldn't be out of place in military boot camp, as described in the next chapter.

---

* In 2009, NASA, the ESA, and the Japanese space agency sent observers to watch how Canada selects its astronauts.

3

Selection

THE TRIALS AND TRIBULATIONS OF BEING SELECTED FOR THE BEST JOB ON AND OFF THE EARTH

It's extremely difficult to be selected as a NASA astronaut, but if you happen to be a Canadian or a European, the odds are almost non-existent. Whereas the Americans have the luxury of having a space agency that announces astronaut selection campaigns every four years or so, Canadians and Europeans contend with selections occurring with frustrating infrequency. They also face daunting statistical odds.

In 2008, NASA, the Canadian Space Agency (CSA), and the European Space Agency (ESA) initiated astronaut selection campaigns. For Americans with ambitions to fly to space, the wait for selection had only been four years. For the Europeans and the Canadians, whose last selection had been in 1992, the wait had been considerably longer. Making the challenge of becoming an astronaut even more daunting was the fact the CSA was planning on selecting only two astronauts whereas NASA's call was for 15 astronauts (the American space agency eventually selected nine astronauts in addition to seven Educator Astronauts). Meanwhile, on the other side of the Atlantic, the ESA announced it would select only four candidates (it ended up selecting six). Following the application deadlines, it was possible to calculate the odds of becoming an astronaut, depending on whether you lived in the US, Canada, or Europe. NASA received 3,535 applicants for 16 positions, meaning prospective astronauts faced a 0.4% chance of being selected (or 235 people chasing each astronaut position). The ESA received 8,413 applicants for just four positions, meaning future European spacefarers had a 0.04% chance of being selected (or 2,103 people chasing each astronaut position). However, the applicants with the highest odds against them were the poor Canadians, whose agency received 5,351 applicants for just two positions (or 2,676 people chasing each astronaut position!).

NASA SELECTION

Following an initial review of 3,535 applicants, hundreds of applicants were disqualified based on the application information. Following a more thorough review of the information, 120 applicants were invited to visit Johnson Space Center (JSC) in October, 2008. These applicants were contacted by the Astronaut Selection Office a week before they were scheduled to arrive at JSC. During this phase of selection, two groups of 10 candidates were assessed each week, for a total of six weeks of assessment.

2008 selection process

The first item on the assessment agenda (Table 3.1) was an orientation and welcome by the selection office. This was usually conducted by the chair and deputy chair of the astronaut's selection board, astronauts Peggy Whitson and Steven Lindsay. They explained to the group what it was like to be an astronaut and what the chances of an early demise were! Following the brief overview of the astronaut occupation, candidates were divided into two groups of five and taken on tours of Mission Control, and mockups of the International Space Station (ISS) and the Space Shuttle. Next was a series of anthropometric measurements using a 3D laser measuring system, to make sure each candidate would fit into the Soyuz capsule. Following the anthropometric assessment was a robotics evaluation that determined 3D reasoning skills, situational awareness, and the ability to multitask.

The next day started with a series of written psychological exams, requiring applicants to answer thousands of short questions to assess sociability, teamwork, and a myriad other psychological parameters. After spending five hours answering psychological questions, applicants reported for a medical questionnaire review, after which there was the opportunity for a physical work-out with the Astronaut Strength and Conditioning Team.

The third day included a 1-hr panel interview administered by the astronaut selection board. To introduce themselves to the 12-person board, each applicant wrote down three to five reasons why they wanted to be an astronaut (the Canadian applicants, in contrast, had to write a 1,000-word essay!). The interview usually started with the predictable "Tell us about yourself starting with high school" question, before exploring the themes applicants had indicated in their reasons for becoming an astronaut.

In March, 40 candidates were selected for the next phase of selection, comprising a week-long medical (Table 3.2). Again, the 40 candidates were divided into two groups, with one group performing their medical tests one week and the second group the following week.

Table 3.1. NASA astronaut selection schedule.

Dates	*Selection process*
September, 2007	Vacancy announcement in USAJOBS
July 1st, 2008	Vacancy announcement closes. 3,535 applications received
September to November, 2008	Qualified applicants reviewed to determine Highly Qualified applicants. 450 selected. Qualifications Inquiry form sent to Supervisors/References and civilian applicants contacted by mail to obtain an FAA medical exam
October to January, 2009	120 Highly Qualified applications reviewed to determine interviewees
November, 2008 to January, 2009	Interviewees attend JSC for preliminary interview, medical evaluation, and orientation in groups of 10 for 2.5 days. Interviewees selected from Highly Qualified group and contacted on a week-by-week basis
February, 2009	40 finalists selected
February to April, 2009	Finalists attend JSC for additional interview and medical evaluation (1 week long)
June, 2009	Astronaut Candidate Class of 2009 announced
August, 2009	Astronaut Candidate Class of 2009 reported to JSC

Table 3.2. NASA medical examinations and parameters.

Item	*Description*
1	Medical history
	a. NASA medical survey
	b. Questionnaire
2	Physical examination
	a. General physical
	b. Anthropometry (biometric assessment of the body)
	c. Muscle mass
	d. Pelvic exam and Pap smear
	e. Procto sigmoidoscopy (invasive examination of the large intestine from the rectum through the last part of the colon)
3	Cardio pulmonary evaluation
	a. History and examination
	b. Physical fitness test
	c. Exercise stress test
	d. Blood pressure
	e. Resting and 24-hr electrocardiograph
	f. Echocardiogram (ultrasound technique used to generate 3D image of the heart)
4	Ear, nose and throat (ENT) evaluation
	a. History and examination
	b. Audiometry
	c. Tympanometry (an objective test of middle-ear function. It is not a hearing test, but a measure of energy transmission through the middle ear)
5	Ophthalmological evaluation
	a. Visual acuity, refraction, and accommodation

b. Color and depth perception
c. Phorias (the relative directions of the eyes during binocular fixation on a given object in the absence of an adequate stimulus)
d. Tonometry (procedure ophthalmologists perform to determine the intraocular pressure (IOP) – the fluid pressure inside the eye)
e. Perimetry (systematic measurement of differential light sensitivity in the visual field by the detection of the presence of test targets on a defined background) and retinal photograph
f. Endoscopy

6 Dental examination
a. Panorex (also known as an "orthopantogram", a panorex is a panoramic scanning dental X-ray of the upper and lower jaw showing a 2D view of a half-circle from ear to ear) and full dental X-rays within last 2 years

7 Neurological examination
a. History and examination
b. EEG at rest

8 Psychiatric/psychological evaluation
a. Psychiatric interviews
b. Psychological tests

9 Radiographic evaluation
a. Chest X-ray
b. X-ray DNS
c. Mammography
d. Medical radiation exposure history and interview
e. Abdominal and urogenital ultrasonography (an ultrasound-based diagnostic imaging technique used to visualize subcutaneous body structures)

10 Laboratory investigation
a. Complete hemogram (blood test, including an estimate of the blood hemoglobin level, packed cell volume, and blood count)
b. Blood biochemistry
c. Immunology
d. Serology (scientific study of blood serum. The term usually refers to the diagnostic identification of antibodies in the serum)
e. Endocrinology
f. Urinanalysis
g. 24-hr chemistry
h. Renal stone profile
i. Urine endocrinology
j. Urine RE
k. Stool RE
l. Occult blood
m. Ova and parasites

11 Other tests
a. Drug screen
b. Montoux test (test for tuberculosis)
c. Microbiological, fungal, and viral tests
d. Pregnancy test
e. Screening for STD
f. Abdominal ultrasonography

Figure 3.1. NASA's Class of 2009 Educator Astronauts. Standing left to right: Mike Schmidt, Stephen Heck, Stuart Witt (Mojave Air and Space Port General Manager), Jim Kuhl. Seated left to right: Lanette Oliver, Chantelle Rose, Rachael Manzer, Maureen Adams. Image courtesy: NASA.

Finally, on June 29th, 2009, NASA selected nine astronaut candidates (Table 3.3) and, 3 weeks later, the agency announced an additional seven candidates (Mike Schmidt, Stephen Heck, Jim Kuhl, Lanette Oliver, Chantelle Rose, Rachael Manzer, and Maureen Adams) as part of the Educator Astronaut selection (Figure 3.1).

ESA ASTRONAUT RECRUITMENT CAMPAIGN

The ESA's first astronaut selection was held between 1977 and 1978, following a 1973 agreement the ESA had with NASA to supply the first Spacelab reusable science laboratory. Spacelab was carried in the Space Shuttle's cargo bay in exchange for flight opportunities for European astronauts. The 1977–1978 selection was followed by another campaign in 1991–1992, which received a record 22,000 applications. European astronaut candidates then had to wait 16 years before the ESA's 2008 campaign (Tables 3.4 and 3.5), a selection process that stretched for 10 months.

Table 3.3. NASA's Class of 2009.

Name	*Age*	*Background*
Serena M. Aunon	33	Wyle flight surgeon for NASA's Space Shuttle, ISS, and Constellation Programs. Holds degrees from The George Washington University, University of Texas Health Sciences Center in Houston, and UTMB
Jeanette J. Epps	38	Technical intelligence officer with the CIA. Holds degrees from LeMoyne College and the University of Maryland
Jack D. Fischer	35	Test pilot and US Air Force Strategic Policy intern at the Pentagon. Graduate of the US Air Force Academy and Massachusetts Institute of Technology (MIT)
Michael S. Hopkins	40	Lt. Colonel US Air Force. Special assistant to the Vice Chairman (Joint Chiefs of Staff) at the Pentagon. Holds degrees from the University of Illinois and Stanford University
Kjell N. Lindgren	36	Wyle flight surgeon for NASA's Space Shuttle, ISS, and Constellation Programs. Degrees from the US Air Force Academy, Colorado State University, University of Colorado, the University of Minnesota, and UTMB
Kathleen Rubins	30	Principal investigator and fellow, Whitehead Institute for Biomedical Research at MIT, and conducts research trips to the Congo. Degrees from the University of California–San Diego and Stanford University
Scott D. Tingle	43	Commander US Navy. Test pilot and Assistant Program Manager–Systems Engineering at Naval Air Station Patuxent River. Degrees from Southeastern Massachusetts University and Purdue University
Mark T. Vande Hei	42	Lt. Colonel US Army. Flight controller for the ISS at NASA's Johnson Space Center, as part of US Army NASA Detachment. Graduate of Saint John's University and Stanford University
Gregory R. Wiseman	33	Lt. Commander US Navy. Test pilot. Department Head, Strike Fighter Squadron 103, USS Dwight D. Eisenhower. Graduate of Rensselaer Polytechnic Institute and Johns Hopkins University

Table 3.4. The ESA's astronaut selection schedule.

Dates	*Selection process*
May 19th to June 18th, 2008	Application period with a valid medical certificate
July to August, 2008	First step of psychological testing (Hamburg) (1 day for each applicant)
September to December, 2008	Second step of psychological testing (Cologne) (1 day for each applicant)
January to February, 2009	Medical testing (Toulouse or Cologne) (5 days for each applicant)
April/May, 2009	Decision and appointment by Director General, ESA
September, 2009	Start of basic training

Table 3.5. Total number of ESA applicants at the closure of the application period.

Country	*No. of applicants*	*Percent of total applicants*	*As 2nd citizenship*[1]	*Men*[2]	*Women*[2]
Austria	210	2.5	8	195	23
Belgium	253	3.0	8	224	37
Denmark	35	0.4	4	34	5
Finland	336	4.0	5	283	58
France	1,860	22.1	58	1,616	302
Germany	1,798	21.4	35	1,523	310
Greece	159	1.9	14	152	21
Ireland	128	1.5	11	110	29
Italy	927	11.0	39	815	151
Luxembourg	14	0.2	0	14	0
Norway	74	0.9	2	67	9
Other	72	0.9	309	301	80
Portugal	210	2.5	10	192	28
Spain	789	9.4	21	707	103
Sweden	172	2.0	9	156	25
Switzerland	351	4.2	26	325	52
The Netherlands	203	2.4	2	175	30
United Kingdom	822	9.8	42	697	167
Total	*8,413*	*100.0*	*603*	*7,586*	*1,430*

[1] Number of applicants having stated this country as their second citizenship.
[2] Number of applicants being a citizen of this country either as unique or second citizenship.

The selection process

One of the first goals of the prospective applicants was to obtain a JAR-FCL 3, Class 2 medical examination. The JAR-FCL 3 Class 2 was selected by the ESA as a medical examination capable of efficiently and relatively inexpensively detecting many of the most common health-related factors preventing an applicant from becoming a private pilot, and hence also an astronaut. The choice of the JAR-FCL 3 (Panel 3.1 and Table 3.6) also streamlined the selection process and ensured applicants with the most likely chance for success were approved to continue to the next phase of the selection process.

Panel 3.1. JAR-FCL 3 Class 2 medical

The JAR-FCL 3 Class 2 medical certificate is a European-wide accepted standard developed by the Joint Aviation Authority (JAA). It can only be issued by specifically certified aeromedical examiners. Recognizing it might not have been feasible for each applicant to obtain a JAR-FCL 3 Class 2 medical certificate, the ESA also accepted an equivalent medical statement that had been authorized by a physician.

Table 3.6. ESA medical examinations and parameters.

Item	*Description*
1	Interview/questionnaire evaluation of family history, personal history, and medical history
2	General physical examination, including: a. all major organ systems, including skin; b. mobility of extremities, joints and spine; c. routine ears–nose–throat examination: i. Hearing analysis applicant must be able to understand correctly ordinary conversational speech at a distance of 2 m from and with his back turned towards the examiner; d. basic neurology assessment; e. genito-urinary evaluation; f. for females: gynecologic evaluation; g. resting heart rate and blood pressure.
3	Standard 12-lead resting ECG
4	Blood analysis for hemoglobin (Hb), lipids and cholesterol
5	Urine stick analysis, including glucose, leucocytes, erythrocytes, protein
6	Ophthalmologic analysis: a. vision in both eyes; b. distant visual acuity with or without correction of 0.5 (6/12) or better in each eye separately and with both eyes 1.0 (6/6) or better: i. Refractive error shall not exceed +5 to –8 diopters; ii. Astigmatism shall not exceed 3 diopters; c. normal color perception (Ishihara or Nagel's anomaloscope).

Following the first round of evaluation, the ESA selected 918 candidates for computer-based psychological testing. From this first batch of 918, the ESA selected 192 candidates. These candidates were invited to the second stage of psychological testing, which commenced in the European Astronaut Center (EAC), Cologne, Germany, in September, and continued until December. While at EAC, astronaut candidates were evaluated by more psychological testing and were assessed in role-playing exercises. They also conducted computer simulations and were interviewed. From the 192 hopefuls, the ESA selected 80 to return for extensive medical evaluation, which took place in January and February. Following this stage, 40 candidates were invited for a formal panel interview from which six candidates (Table 3.7) were selected (the ESA had originally announced it would select four astronauts) in May, 2009 (Figure 3.2).

Figure 3.2. The ESA's Astronaut Class of 2009. The six individuals who will become Europe's new astronauts were presented at a press conference held at ESA Headquarters in Paris, France, on May 20th, 2009. From left to right: Luca Parmitano (It), Alexander Gerst (Ger), Andreas Mogensen (Den), Samantha Cristoforetti (It), Timothy Peake (UK), and Thomas Pesquet (Fr). Image courtesy: ESA – S. Corvaja, 2009.

Table 3.7. The six candidates selected after formal panel interview.

Name	*Age*	
Samantha Cristoforetti (Italian)	32	Master's degrees in engineering and in aeronautical sciences from the University of Naples Federico II in Italy. Works as a fighter pilot with the Italian Air Force
Alexander Gerst (German)	33	Diploma in geophysics. Studied Earth science at Victoria University of Wellington in New Zealand, where he was awarded a Master of Science. Working as a researcher since 2001
Andreas Mogensen (Danish)	33	Master's degree in engineering from Imperial College, London, and a doctorate in engineering from the University of Texas. Worked as an attitude and orbit control system and guidance, navigation and control engineer for HE Space Operations
Luca Parmitano (Italian)	33	Holds a diploma in aeronautical sciences from the Italian Air Force Academy. Trained as a Full Experimental Test Pilot at EPNER, the French test pilot school in Istres. Pilot with the Italian Air Force
Timothy Peake (British)	37	Degree in flight dynamics and qualified as a Full Experimental Test Pilot at the UK's Empire Test Pilots' School. Officer serving with Her Majesty's Forces as an Experimental Test Pilot
Thomas Pesquet (French)	31	Master's degree from the Ecole Nationale Supérieure de l'Aéronautique et de l'Espace in Toulouse, France. Worked at the French space agency, CNES, as a research engineer. Currently flies Airbus A320s for Air France

CANADIAN SPACE AGENCY

Canada's third astronaut recruitment campaign (Table 3.8) was by far the most challenging of any astronaut selection to date. Not only did applicants face daunting statistical odds, but the selection process subjected them to a series of rigorous and grueling evaluations, many of which had applicants feeling they were in boot camp! After the cull following the preliminary interview, 40 applicants (Table 3.9) were selected to travel to Saint Hubert, Quebec, home of the CSA, for a series of tests ranging from "flying" the Canadarm2 to underwater problem-solving.

Saint-Hubert

The Saint-Hubert testing phase was divided into two groups, one group arriving on a Saturday evening and the other on the Sunday. The first evening was a social event featuring an introduction by Jean Marc Comtois, Head of Operational Space Medicine, who explained the significance of the selection and outlined what tests and

Table 3.8. Canadian Space Agency astronaut selection schedule.

Dates	*Selection process*
May, 2008	Campaign launch
	Application deadline. 5,351 applications received
October to December, 2008	Preliminary interviews. 79 candidates
January, 2009	Robotics, flight operations, and physical tests. 39 candidates
February, 2009	Survival and damage control tests. 31 candidates
March, 2009	Medical and psychological tests. 16 candidates (average age: 36 years)
May, 2009	Announcement of two successful candidates
June, 2009	Astronauts report to Saint Hubert, Quebec, for duty
August, 2009	Astronauts report to JSC for training with NASA's 2009 Astronaut Class

Table 3.9. Breakdown of the top 40 astronaut candidates by region.

Region	*Candidates*	*Region*	*Candidates*
Atlantic Provinces	1	Western (British Columbia)	4
Central (Quebec)	8	Northern Canada	0
Central (Ontario)	10	Outside Canada[1]	11
Prairies	6		

[1] Canadian citizens residing outside the country.

assessments we had to look forward to. More than a quarter of the candidates at this stage were military pilots, with medical doctors coming in a close second. The rest of the candidates were a mix of over-achieving scientists and CSA employees, some of whom had worked with Jean Marc. Although everyone was friendly, some were already sizing up the competition, wondering if they measured up and what their weaknesses might be. Some of those weaknesses became apparent during the following days of testing.

St. Jean

The first day was spent at St. Jean, home to the Canadian Forces CF's basic training base. Here, we were subjected to myriad exercise tests and assessments, beginning with the CF's basic fitness test with a few twists.

The first test was the multi-stage fitness test (MSFT) (Panel 3.2), or "bleep test", familiar to everyone with a military background. After the MSFT (Figure 3.3), we moved on to press-ups, sit-ups, chin-ups, and some uncomfortable isometric tests. Throughout the tests, we were under the supervision of CF Personnel Support Agency (PSP) staff who treated some of the most highly trained humans on the planet as if they were day-one, week-one cadets. No one was allowed to talk or look over their shoulder to see how candidates were performing in the individual tests. For those in the military, it was just another day at the office, but for the civilians, it was a little strange.

Panel 3.2. Multi-stage fitness test

The MSFT is used by sports coaches to estimate maximum oxygen uptake. The test involves running continuously between two points 20 m apart. These shuttle runs are synchronized with a pre-recorded audio tape, which plays beeps at set intervals. As the test proceeds, the interval between each successive beep reduces, forcing the victim to increase speed until he/she is incapable of keeping up with the recording. The recording is structured into 21 "levels", each lasting around 63 sec. The interval of beeps is calculated as requiring a speed at the start of 8.0 km/h, increasing by 0.5 km/h with each level. The progression from one level to the next is signaled by three rapid beeps. The highest level attained before failing to keep up is recorded as the score for that test.

Figure 3.3. Canadian Space Agency astronaut applicants performing the Multi-Stage Fitness Test at St. Jean, Quebec, in January, 2009. At this phase of the selection, 39 applicants remained in the running! Image courtesy: CSA.

Next was a series of swimming and sea survival tests. In the initial application form, there had been a question that asked whether the applicant was comfortable with water. I would imagine most people would answer that in the affirmative, since you have to drink it every day to stay alive! What the question should have asked was whether the candidate could swim! Because of this slight administrative error, there were two candidates who couldn't swim! Since this phase of selection was focused upon select-in criteria, the two non-swimmers were guaranteed not to progress to the next stage.

The first swim test was a very leisurely 250 m in 10 min. This was followed by a dive into the pool from 3 m and treading water for 10 min with our hands above the surface. Then, we had to jump into the water from the 5-m board and swim to the side wearing a life preserver before conducting the underwater "shapes" task. The "shapes" test required each candidate to dive to the bottom of the pool, wearing a weight belt for ballast, and fit differently shaped blocks of wood into a container. Some candidates failed to place one block into the container, while one candidate, who was capable of holding his breath for 5 min, managed to fit all the blocks and then wondered whether he would be penalized for taking so long!

Aircrew selection and personality testing

The second day of testing was conducted at the CSA's Saint Hubert location, where we spent one morning completing the CF's aircrew selection test (Figure 3.4) – a

Figure 3.4. CSA astronaut applicants, Kenneth Welch (left) and Bruce Woodley (right), perform the Canadian Force's aircrew selection tests at Saint-Hubert, Quebec, in January, 2009. Image courtesy: CSA (*see colour section*)..

series of more than 20 computer-based tests designed to assess cognitive, spatial, and motor skills. Needless to say, the test pilots didn't find this testing phase particularly demanding!

Next, we performed a series of hand–eye coordination tests on paper, followed by a public service exam normally written by diplomats serving abroad. Every once in a while, in between testing sessions, one of us would be asked to conduct a media interview in front of television cameras. After the 1-hr public service exam, we answered more than 1,200 questions in three separate psychology questionnaires, one of them being the Minnesota Multiphasic Personality Inventory (MMPI), one of the most frequently used personality tests in mental health (used to identify personality structure and psychopathology). The MMPI alone had 567 questions, which meant by the time they handed us the second questionnaire, our hands were cramping badly.

Robotics assessment

The third day of testing focused primarily on robotics. Unlike the 2-week course offered to astronauts during their advanced and increment-specific training, we received just 2 hr of instruction on how to manipulate the Canadarm2 (Panel 3.3), using scale models of the arm and latching end effectors (LEEs), before being assessed by a robotics instructor (Figure 3.5).

Figure 3.5. CSA astronaut applicant, Chris Denny, "flies" the Canadarm2 during the robotics evaluation. Image courtesy: CSA.

Panel 3.3. Challenges of manipulating the robotic arm

Manipulating the arm is achieved by operating two Robotic Work Station (RWS) control joysticks while looking at the RWS screens shown in Figure 3.5. One joystick is the Rotational Hand Controller (RHC) and the other is the Translational Hand Controller (THC). The screens you can see in Figure 3.5 are the Display and Control Panel (DCP) and the Portable Computer System (PCS) laptop, which provide the operator with a view similar to what an astronaut would see in the ISS.

The most challenging aspect of operation was understanding the coordinate frames. During operations, the operator must understand along which axis the LEE (and attached payload if there is one) will move, and around which point in space it will rotate. To do this, several coordinate frames are used to generate digital position and attitude displays. For example, fundamental elements are the Frame of Resolution (FOR), the Display Frame, and the Command Frame. The FOR defines the manipulator or attached payload multidimensional position (x, y, z) and attitude (pitch, yaw, roll). The Display Frame is the reference coordinate frame for the FOR to compute and display the position and attitude. The Command Frame determines the direction of motion of the arm/attached payload.

The selection of desired FOR, Display Frame, and Command Frame is a contributing factor in determining the degree of difficulty of a robotics task. The position of the manipulator or attached payload with respect to a base structure, vector of arm maneuver, and available visual cues are the major factors in determining the optimum combination of these coordinate frames. As you can see in Figure 3.5, our situational awareness was dependent on cameras and derived digital information sources. Although our assessed task was a simple (for a trained astronaut) capture task, in reality, field of view, reference frames, and dynamically changing conditions make high demands on the operator's ability to comprehend current status and determine implications of the next control input.

The robotics assessment brought to a close the Saint-Hubert phase of testing. Although we weren't told what the next phase of testing would involve, we were told it would take place in Halifax, Nova Scotia. For those with a military background, that could only mean damage control and sea survival (CF helicopter pilots visit Halifax to perform their dunking and sea survival tests).

Halifax

A month later, 31 candidates were in Halifax for the penultimate stage of selection. The first day was spent at the Navy's Damage Control School, a high-fidelity

Figure 3.6. CSA astronaut applicants being assessed during a hazardous material exercise at the Canadian Forces Damage Control School in Halifax, Nova Scotia, in February, 2009. Image courtesy: CSA (*see colour section*).

mockup of various parts of a ship. Inside the Damage Control School, sailors are taught how to fight fires and flood in addition to learning how to react to hazardous material (HazMat) spills.

Damage control school

Before lunch on the first day, we had already been given a demonstration of the fire rooms and flood tanks and been fitted with bunker gear. After lunch, my group was assessed on a HazMat spill, requiring us to suit up in blue Tyvek bunny suits (Figure 3.6) and enter a room in which a toxic substance had spilled. Among the tasks was evacuating an injured person and securing the room while maintaining communication with command. During the debrief following the assessment, we were told we had executed procedures correctly, but had taken a long time to secure the room. Given that our instruction had consisted of a 15-min Powerpoint presentation, my group figured we hadn't performed too badly.

Fighting floods

The next day began with a flood exercise requiring us to suit up in coveralls and wetsuits and enter a compartment slowly filling with icy water. Peering at us through

Figure 3.7. A CSA astronaut applicant, Chris Denny, constructs a strongback while being assessed during a damage control exercise. Image courtesy: CSA.

an observation window were CF damage control instructors and members of the CSA's astronaut selection group. The exercise was an opportunity for us to show how well (or badly!) we worked as team in a stressful environment. The exercise started out promisingly enough as we barked orders and quickly assigned team roles to one another. Then we prioritized the leaks and started to work on stopping the ingress of water. Due to the number of leaks, we split into two teams, one of us working on the smaller leaks and the other cutting shoring to construct a strongback (Figure 3.7). While this was going on, hundreds of liters of cold water continued to pour in. By the time we had plugged the first major leak, we were up to our chests in cold water and had lost one team member to hypothermia. Eventually, we plugged most of the leaks before the instructors – mercifully – called an end to the exercise.

We were expecting a debrief but there wasn't time, as our instructor rushed us off to the fire section, where we were told to don bunker gear. Less than 5 min later, we were dressed in bunker gear holding a fire hose (Figure 3.8) and led into a dark room that normally doubled for an engineering space onboard a ship. We couldn't see a thing until a fire started in the corner and we turned the hose on. In seconds, the fire had spread across the room and was creeping across the ceiling. At least we were warm! After half an hour of fire-fighting, it was back to the flood area, where we had to don coveralls again and plug more holes. A break for lunch was followed by a casualty search exercise in an inter-connected series of smoke-filled compartments. Dressed in bunker gear, we had to find our way inch by inch using a thermal imaging camera (TIC), while checking for casualties and rendering assistance. Every once in a while, the instructors would turn the flames on and we'd have to make a quick assessment of what type of fire it was before searching for the right fire extinguisher (for some reason, these were never readily available and we would have to backtrack

Figure 3.8. CSA astronaut applicants struggle to patch a fire main during an assessed team-building exercise at the Canadian Forces Damage Control School in February, 2009. Image courtesy: CSA.

to search for them) and extinguishing the fire. After the second fire exercise, we thought that might be the end of the fire and flood exercises, but the instructors wanted to see us get soaked one more time so they "exploded" an overhead fire main on us (Figure 3.8). With some quick thinking, we managed to run a bypass, which slowed the torrent of water to a trickle, but only for a few seconds. Seeing we had solved their problem too quickly, the instructors burst another main and we were drenched again. Then, as soon as we started patching the hole, the instructors took one of our group away for another problem-solving exercise. We spent at least 30 min patching before we finally managed to patch the fire main, by which time we were all hypothermic once again, so we didn't look our best when the television crew arrived!

Survival Systems

The final day of testing was held at Survival Systems in Halifax, where we were taken in one at a time to be evaluated in the helo-dunker (Figure 3.9). Once in the mock helicopter cockpit, we were strapped in and given escape instructions before we were sent crashing into the swimming pool. The dunker then rolled over and quickly filled with water. An instructor then tapped us on the shoulder, which was the signal for us to escape. We did this three times, each scenario being more complicated than the last. Finally, after surviving the dunking, we were led to the top of a 10-m tower and told to jump into the swimming pool and clamber into a life-raft. It was easily the most enjoyable day of selection.

Figure 3.9. CSA astronaut applicants being evaluated inside Survival Systems' helo-dunker. Image courtesy: CSA (*see colour section*).

The final cut

After being exposed to fire, flood, and hypothermia, the candidates returned home and waited for the e-mail informing them they had progressed to the final 16. The final cut to 16 eliminated some of the best candidates, many of whom would probably have been selected as astronauts if only they had had an American passport. In fact, several of those working in the US who missed the final cut were already making plans to obtain a Green Card so they would be eligible for NASA's next selection; patriotism takes a distant second place when you've spent your whole life wanting to fly in space!

Sixteen candidates

The final 16 were introduced at a press conference at the Defense Research and Development facility in Toronto in March. After the media event, there followed a week of various medical procedures ranging from brain magnetic resonance imaging (MRI) to 3D anthropometric assessment (Figure 3.10).

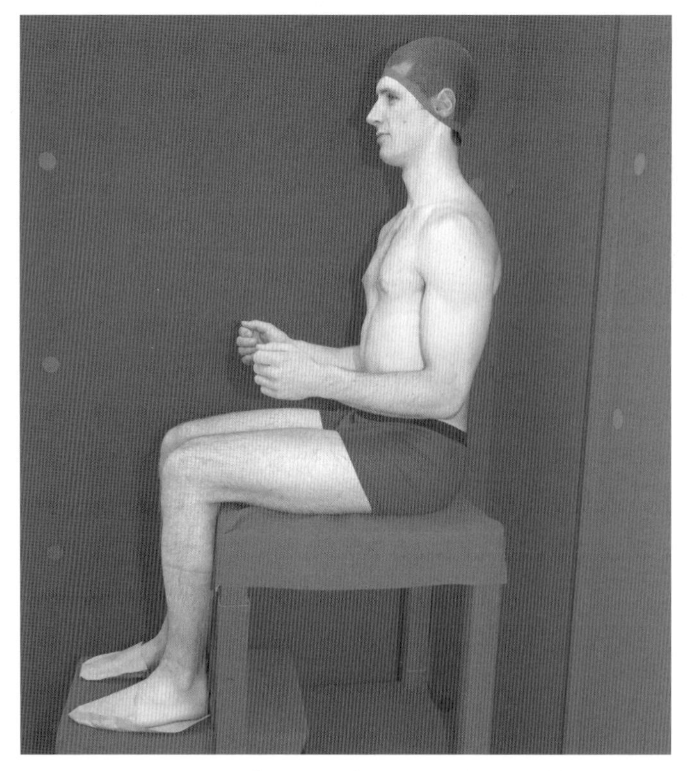

Figure 3.10. CSA astronaut applicant, Joshua Kortryk, sits while a 3D anthropometric assessment is conducted during the final round of selection. Image courtesy: CSA.

Figure 3.11. New Canadian astronaut, David Saint-Jacques (left), Industry Minister Tony Clement, and new Canadian astronaut, Jeremy Hansen (right). Image courtesy: CSA.

Selection

> "It takes so many things to line up and I was worried I might be applying a little early. If I don't make it here and they do another search in 16 years, I hope I'll still be in good physical condition when I'm 42 to apply again."
>
> Joshua Kutryk, final 16 candidate and fighter pilot with the CF

After a job interview lasting the better part of a year, two candidates made the final grade: Dr. David Saint-Jacques and Captain Jeremy Hansen (Table 3.10 and Figure 3.11). As Canada's two newest astronauts, they have the exciting prospect of not only following in the footsteps of the likes of Bob Thirsk and Julie Payette, but also of furthering Canada's role on the ISS and missions to the Moon. Before they have these opportunities, though, they face a lengthy period of probationary training, technical assignments, and mission training.

Table 3.10. The CSA's Class of 2009.

Name	*Age*	*Background*
Jeremy Hansen	33	Bachelor of Science degree in Space Science from Royal Military College in Kingston, Ontario. Master of Science in Physics from the same institution in 2000. Prior to joining the Canadian Space Program, Capt. Hansen served as a CF-18 fighter pilot and held the position of Combat Operations Officer at Four Wing Operations in Cold Lake, Alberta
David Saint-Jacques	39	Bachelor of Engineering degree in Engineering Physics from École Polytechnique de Montréal (1993). Ph.D. in Astrophysics from Cambridge University, UK (1998). MD from Université Laval in Quebec City, Quebec (2005). Completed family medicine residency at McGill University (2007). Prior to joining the Canadian Space Program, Dr. Saint-Jacques was a medical doctor and Co-chief of Medicine at Inuulitsivik Health Centre in Puvirnituq, Quebec

Section II

Preparing for Life in Space

Before they have their opportunity to fly in space, astronauts undergo hundreds of hours of training, which is divided into three phases. First, the newly selected astronaut candidates (ascans) must pass a course of probationary training. During this phase, the ascans learn about space technology, basic medical skills, and how the International Space Station (ISS) works. They also become familiar with scuba-diving. After the first phase, they go on to more advanced training, where they learn in more detail about the various parts of the ISS, the experiments, the spacecraft, and the involvement of ground control. Following probationary training, astronauts must then spend time performing technical assignments before eventually being assigned to a mission, which requires more training! During mission training, the astronauts work as much as possible with the other members of the crew, they learn about the special tasks linked with their mission, and they become familiar with weightlessness doing parabolic flights. They also learn a foreign language, learn about the scientific experiments, and visit training centers in the US, Russia, Japan, Canada, and Europe.

4

Astronaut probationary training

INTERNATIONAL SPACE STATION TRAINING FLOW

The multilaterally agreed training approach for International Space Station (ISS) increments consists of three consecutive phases: Basic Training, Advanced Training, and Increment-Specific Training (Figure 4.1). The Basic Training phase, which takes a year, is provided independently by each ISS International Partner for its astronauts. However, some agencies, such as the Canadian Space Agency (CSA),

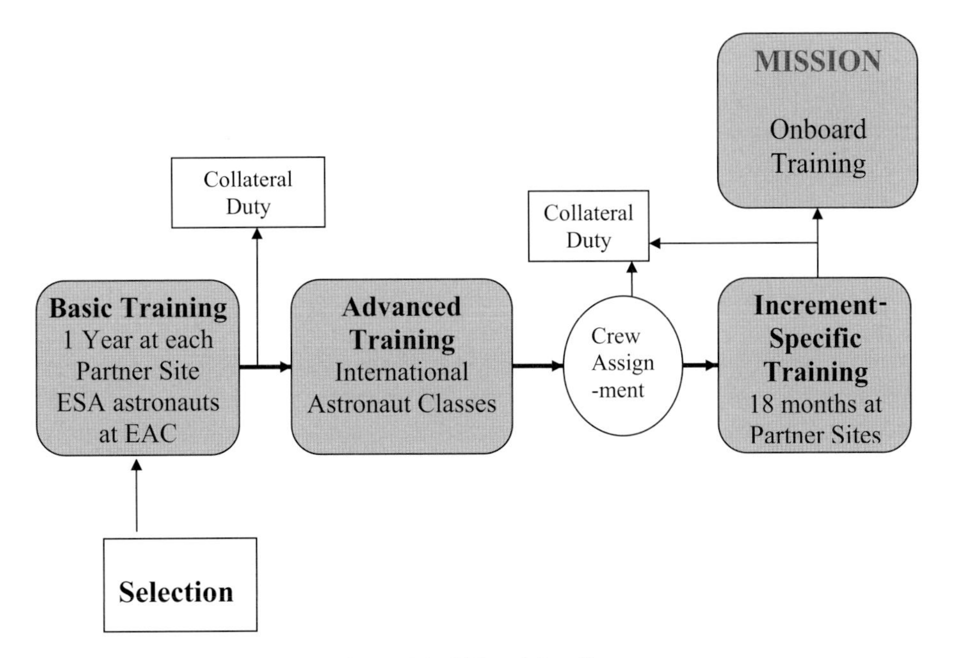

Figure 4.1. ISS training flow.

send their astronauts to be trained by NASA for this phase. For example, Jeremy Hansen and David Saint-Jacques, the two CSA astronauts selected in 2009, trained with NASA's 2009 Class. The following two training phases are designed for international training classes and are conducted at International Partner training sites. During these multilateral training phases, each Partner is responsible for providing training to all ISS astronauts for those elements it is contributing to the ISS Program.

Basic Training

The first training phase for newly recruited astronauts (which lasts between 12 and 16 months, depending on which agency is conducting the training) provides the basic knowledge required by a professional astronaut. Basic Training includes an introduction to space organizations and current programs, spaceflight fundamentals such as engineering and life sciences, and space systems and operations associated with the ISS. This phase also fosters the development of basic skills considered necessary for astronauts in the Advanced Training phase, such as scuba-diving and public relations activities.

Advanced Training

This phase of training also lasts a year and provides a more thorough understanding of ISS systems and subsystems, payloads, and launch vehicles. It focuses primarily on generic ISS onboard tasks and the interactions with ground centers. This phase also prepares astronauts for their Increment-Specific Training (IST) and their first flight assignment. For example, the Advanced Training includes some specialization such as robotics and extravehicular activity (EVA). Once astronauts have completed this phase, they are eligible for mission assignment.

Increment-Specific Training

This final training phase lasts 18 months. Its primary objective is to prepare the prime and backup crews assigned to an ISS flight to perform all the activities planned for that particular flight increment. This includes standard operations and maintenance tasks for systems and payloads, as well as activation and checkout of new ISS elements.

This chapter focuses on the first two phases of training for NASA and ESA astronauts and describes how astronaut candidates (ascans) become astronauts in a training period stretching over 2 years.

EUROPEAN SPACE AGENCY

Of all the astronaut training programs, the ESA's is perhaps not only the most challenging to follow, but also the most challenging to organize. This is because the decentralized training provided by the International Partners participating in the ISS program demands a high degree of coordination and synchronization. Furthermore, the teaching methods, training concepts, *and* the multicultural backgrounds of the astronaut candidates and their instructors must be standardized.

Integrated training schedule

Once the training has been coordinated and synchronized, the training aspects must then fit into one overall integrated schedule. This is a challenge in itself, since each astronaut has an individually tailored training plan, meaning no two astronauts get exactly the same training at the same time. If you consider between 30 and 40 astronauts/cosmonauts undergo training in one year at five different sites, you get some idea of the tremendous organizational effort required. Fortunately, the ISS partner space agencies have successfully managed to synchronize the training for more than a decade now.

Phases of astronaut training

Each ESA astronaut starts the astronaut training cycle by first completing the 16-month Basic Training at the European Astronaut Centre (EAC) in Cologne (Figure 4.2), Germany. This phase, which provides the astronaut candidate with a solid introduction to their future career as an astronaut, consists of four training blocks:

1. Introduction
2. Fundamentals
3. Space Systems and Operations
4. Special Skills.

1. Introduction

This phase is a little like the first day at work, since it provides the new hires a broad overview of the arena in which they will be working for the next 10–20 years. For example, they are introduced to the policies of the major spacefaring nations, their space agencies (with special emphasis on the ESA!) as well as the major manned and unmanned space programs. Spicing up this phase are lectures covering the basics of space law and the intergovernmental agreements governing worldwide cooperation in space.

Figure 4.2. The European Space Agency's Astronaut Training Center, in Cologne, Germany. Image courtesy: ESA.

2. Fundamentals

The second phase of basic training teaches the candidates basic knowledge on various technical and scientific disciplines. The objective of this phase is to ensure all new astronaut candidates, who have different professional backgrounds and expertise, have a common minimum knowledge base in subjects relevant to their career. Although the phase deals only with the fundamentals, an astronaut's job requires knowledge of so many different subjects that candidates often feel as if they're drinking from a fire-hose as a result of all the information delivered to them by their instructors. In addition to covering technical disciplines such as spaceflight engineering, electrical engineering, aerodynamics, propulsion, and orbital mechanics, this phase also includes an introduction to science disciplines such as research into weightlessness (in human physiology, biology, and material sciences), Earth observation, and astronomy.

3. Space Systems and Operations

This phase provides candidates with a detailed overview of all ISS systems such as guidance navigation & control (GNC), thermal control, electrical power generation

Figure 4.3. Soyuz launcher. Image courtesy: NASA.

and distribution, command and tracking, and life support systems. It also introduces the ascans to robotic, EVA, and payload systems, in addition to the major systems of those spacecraft which service the ISS such as the Russian Soyuz (Figure 4.3) and Progress spacecraft. Additionally, candidates become acquainted with ground systems such as development and test sites, launch sites, and training and control centers.

4. Special Skills

The Special Skills phase focuses on developing skills used for generic robotic operations, rendezvous, and docking, Russian language, human behavior and performance, and scuba-diving. Scuba-diving is an essential skill for those astronauts who will eventually perform EVAs. The Special Skills phase completes the basic training of ESA astronaut candidates and marks the beginning of the advanced training increment.

Advanced training

During the advanced training phase, ascans learn how to service and operate the different ISS modules, systems, and subsystems, and to fly and dock transport vehicles like the Russian Soyuz vessel or the ESA's Automated Transfer Vehicle (ATV, Figure 4.4).

Figure 4.4. *Jules Verne* Automated Transfer Vehicle approaches the International Space Station on Monday, March 31st, 2008. Image courtesy: ESA.

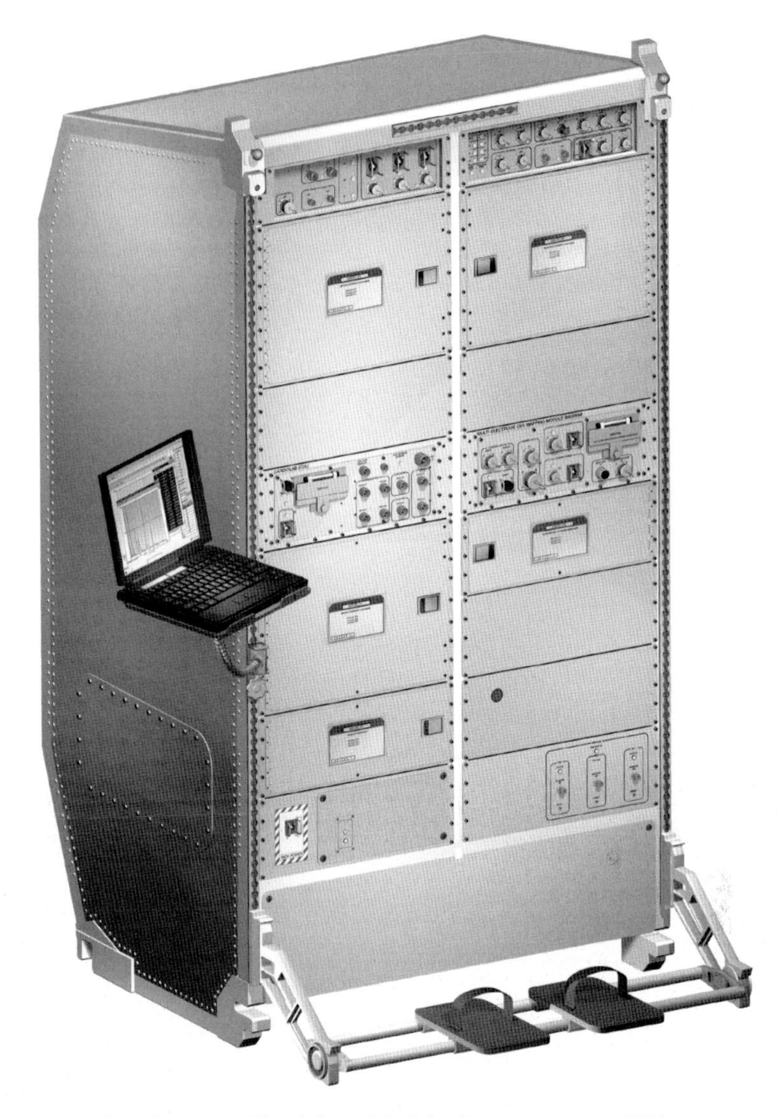

Figure 4.5. European Physiology Module. Image courtesy: ESA.

They also learn how to perform scientific experiments in the ESA's ISS research module, *Columbus* (Panel 4.1), and how to use the equipment fitted in the module's payload racks, such as the European Physiology Module facility (EPM, Figure 4.5).

Panel 4.1. *Columbus* laboratory

The *Columbus* laboratory is the ESA's most significant contribution to the ISS. The 4.5-m diameter cylindrical module is equipped with flexible research facilities providing extensive science capabilities. The laboratory has room for 10 International Standard Payload Racks (ISPRs), each the size of a telephone booth, and is capable of hosting its own autonomous and independent laboratory, complete with power and cooling systems, and video and data links back to researchers on Earth. The ESA has developed a range of payload racks, each designed to maximize the amount of research, and offers European scientists full access to a weightless environment.

At launch, *Columbus* was fitted with five internal payload racks:

a. Biolab supports experiments on micro-organisms, cells and tissue cultures, and small plants.
b. The European Physiology Modules Facility (EPM) enables investigation into the effects of long-duration spaceflight on the human body.
c. The Fluid Science Laboratory (FSL) accommodates experiments in the behavior of weightless liquids.
d. The European Drawer Rack (EDR) is a modular experiment carrier system for a variety of scientific disciplines and provides basic resources for experiment modules housed within standardized drawers and lockers.
e. The European Transport Carrier (ETC) accommodates items for transport and stowage.

Since scientists on the ground rely on astronauts to conduct their experiments in space, astronauts must be proficient in the operation of the equipment on *Columbus*, which is why training is conducted in the lab during the advanced training phase.

Advanced training takes a year to complete. The training is conducted in international astronaut classes and includes training units at all ISS partner training sites. These training sites are located in Houston, in the US (NASA), Star City near Moscow (Russia), Tsukuba near Tokyo (Japan), Montreal (Canada), and at the EAC in Cologne, Germany. Each of the ISS partners is in charge of providing training to all ISS astronauts on the elements that they contribute to the ISS Program. For example, training at the EAC focuses on the *Columbus* module, the ATV, and ESA science training.

It is impossible to describe all the training elements of the ESA's astronaut training program in detail. Instead, to give the reader an idea of the nature of the training, the following section describes EVA training, which constitutes a key advanced training increment.

EVA Pre-Familiarization Training

The ESA's EAC developed the EVA Pre-Familiarization Training Programme (EPFTP) to bridge the gap between scuba training (which astronauts receive during basic training) and NASA's EVA skills training. By completing the EPFTP, astronauts develop the essential cognitive, psychomotor, and behavioral skills required to prepare them to use the ISS spacesuit in NASA's Neutral Buoyancy Laboratory (NBL). Psychomotor skills are required to move in the cumbersome spacesuit and move along the handrails of the ISS. Cognitive skills are important to navigate around the space station and to apply tethering and operational rules, while behavioral skills are useful for spatial and situational awareness, and problem solving.

The EPFTP teaches astronauts to acquire these skills in a 2-week course comprising a series of classroom courses and in-water exercises designed to challenge the astronauts while performing simulated EVAs. First, astronauts are provided with an overview of the EMU suit, including its biomechanics and constraints in water and space. Next, instructors give astronauts recommendations on the best strategies for moving in the suit without fighting it and also the no-go zones on the space station. Next, in a simulated EVA session supervised by instructors, astronauts have the opportunity to test their movement strategies under water (Figure 4.6) in the

Figure 4.6. The ESA's EVA pre-familiarization training takes place at the EAC, in Cologne. Image courtesy: ESA.

EAC's Neutral Buoyancy Facility (NBF, Table 4.1). In this session, the astronauts practice different body postures, changing attitude and body orientation around the confining structures of the underwater station. After their underwater EVA indoctrination session, the astronauts are introduced to the tools they will be using outside the ISS and the interfaces to equipment that will be used in future exercises. After becoming familiar with their EVA tools, astronauts must then be qualified in the Surface Supplied Diving System (SSDS), enabling them to communicate with the Control Room during their underwater training.

Table 4.1. The ESA's Neutral Buoyancy Facility characteristics.

Water tank				
Length	Width	Depth	Volume	Temperature
22 m	17 m	10 m	3,747 m^3	27–29°C

Scuba and SSDS diving equipment

20 complete sets of scuba-diving equipment (cylinders, regulators, suits, etc.)
Three SSDS sets comprising:

- Full-face mask with microphone and earphones for two-way communication between SSDS divers and on-deck personnel
- Buoyancy jacket including inflator, with 6-l/300-bar reserve air tank, pressure gauge, and dive computer
- 60-m umbilical hose connected to deck air supply and for communication cables

SSDS cart on-deck hosting umbilicals, air tanks, pressure-monitoring devices, and video and audio monitoring

EVA tools	
EVA connectors (electrical and fluid)	Limited sets of EVA and dummy tethers
Portable Foot Restraints (PFRs) mounted on EVA worksites on *Columbus* mockup	ISS handrails mounted on airlock and *Columbus* mockup
EMU-like boots for use with PFR	EMU PLSS backpack, helmet, and gloves

NBF control room	
Video: eight channels, including switching matrix for observation and multiple recording of underwater and deck operations	Audio: two audio loops for bi-directional communications between deck personnel and between deck personnel and divers
Additional monitor with switching matrix for deck personnel in NBF hall	Underwater loudspeakers for unidirectional communication with all divers

After the SSDS qualification, astronauts conduct the "EV1 Run", akin to a real EVA. During EV1, the astronauts wear a low-fidelity mini-work-station strapped to their chests to carry EVA tools, a backpack simulating the EMU's Primary Life Support System (PLSS), and a helmet, in addition to a pair of boots and EMU gloves. Guided by instructors clad in scuba-diving gear, the astronauts perform an end-to-end EVA including airlock egress/ingress, payload transportation, using waist tethers, and operation of ISS connectors. Throughout the session, astronauts must comply with EVA rules such as tethering the body at all times and using only D-rings as tethering points.

The third underwater session is the EV1 + 2 Run. This session is a two-member EVA with an emphasis on team situational awareness, crew communication, and workload management. During this EVA, the team is free to develop their own timeline and decide how the EVA tools are shared. In addition to performing routine tasks, the astronauts must also respond to unexpected equipment failures and unplanned activities, while being controlled by a Test Director. Upon completion of the EPFTP, astronauts are given a study guide and DVD package with the course material, including videos of the EVA runs together with additional reference documentation and EVA skill demonstration videos.

While the ESA's basic training phase is predominantly knowledge-based classroom training, the advanced phase includes several practical elements such as EVA, requiring the ascans to spend time in training mockups and simulators. On completion of the advanced training phase, an ESA ascan is finally eligible for spaceflight assignment, which inevitably means even more training!

NASA

Overview of ascan training

In common with the ESA, much of NASA's basic astronaut training takes place in the classroom, where ascans learn about *Orion* (Figure 4.7) and ISS systems. They also study key disciplines such as Earth sciences, meteorology, space science, and engineering, which will prove helpful in their work in space. Outside the classroom, ascans complete water and land survival training to prepare for an unplanned landing.

Once the basic training period is complete, candidates are qualified to become astronauts. In common with the ESA, the end of basic training is not the end of training, but rather the beginning of the next phase. In the advanced phase, astronauts are grouped with experienced astronauts, who serve as mentors to share knowledge and experience. The goal of the mentoring approach is to ensure each trainee is proficient in all activities related to pre-launch, launch, orbit, entry, and landing. Finally, after completing the advanced phase, astronauts receive their mission assignments, and enter the IST phase, described in Chapter 7.

Figure 4.7. *Orion* mockup. Image courtesy: NASA.

Ascan training week by week

Most (pilots are exempted) ascans spend their first week of training in sunny Pensacola, Florida, where they are trained in the art of water survival and receive instruction on subjects as varied as aviation physiology and ejection techniques. The location of this first phase of astronaut training is Pensacola Naval Air Station, which houses the various simulators and equipment required to ensure ascans learn how to survive in an emergency. The survival training requires each ascan to be scuba qualified and be capable of swimming three lengths of a 25-m pool without stopping, and then swim three lengths of the pool in a flight suit and tennis shoes with no time limit. They must also tread water continuously for 10 min while wearing a flight suit.

Ejection seat training

During ejection seat training (EST), ascans learn proficiency in the essential skills relating to the canopy system, ejection seat preflight checks, and strapping in (Figure

Figure 4.8. NASA Educator Mission Specialist, Joseph M. Acaba, gives an all-clear sign immediately after a simulated aircraft ejection during water survival training at Pensacola Naval Air Station. Image courtesy: NASA (*see colour section*).

4.8). They are also taught the basics of the ejection envelope, ejection decision, optimal body position, ejection initiation, parachute deployment altitudes, malfunctions, and hazards such as flash burn and poor body position.

Emergency egress training

After "riding the rails" of the ejection simulator, ascans move on to emergency egress training (EET). In this part of the training, they learn the drills for bailout during an emergency water egress inside a helo-dunker* (Figure 4.9), which simulates a helicopter crash at sea. The helo egress requires ascans to be strapped into the simulator, before being dropped into the water. Once in the water, the helo sinks and rolls over. The ascan is then tapped on the shoulder by one of the training staff. This is the signal for the ascan to release the straps and escape via designated windows. Once they've mastered the egress, they do it again, only this time wearing a blindfold!

---

* The helo-dunker was one of the tests Canadian Space Agency (CSA) astronaut candidates had to perform during the selection round of the last 30 candidates in the 2008–2009 campaign. Although it sounds a little claustrophobic, the test is actually a lot of fun.

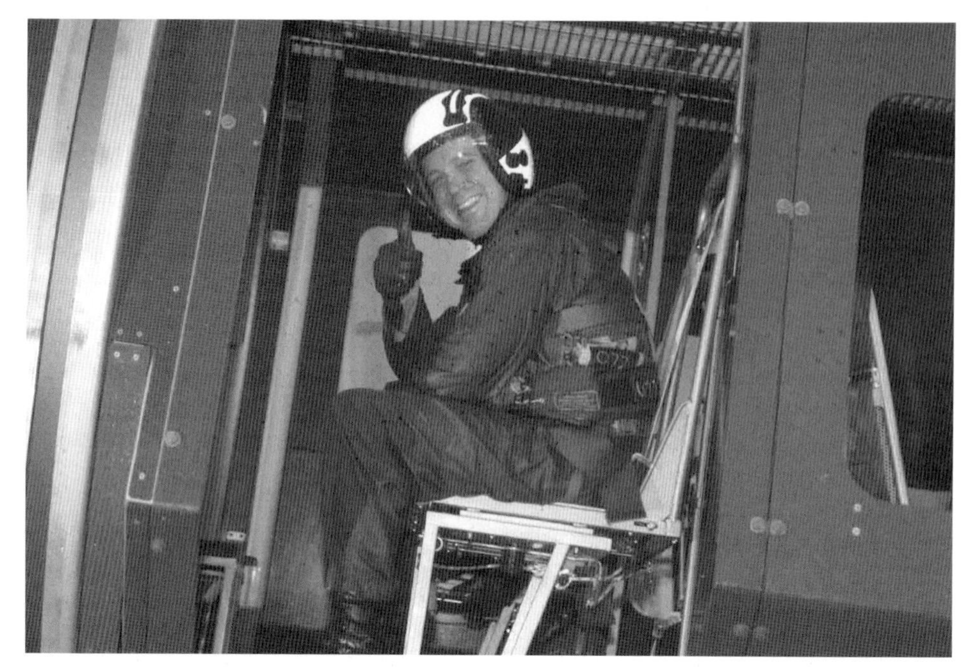

Figure 4.9. One of NASA's Class of 2004 prepares for his helo-dunker indoctrination. Image courtesy: NASA.

Spatial disorientation training

Before being indoctrinated on the Multispatial Disorientation Device (MDD, also known as the "spin and puke" device (!)), ascans receive theory lessons on related topics such as visual illusions, visual scanning, situational awareness, and disorientation countermeasures. After being taught how easy the vestibular system can be tricked, the ascans are subjected to the MDD, an experience teaching them how their eyes and balance can be fooled by darkness and G-forces.

Hypoxia awareness training

After experiencing the MDD, the ascans move on to basic hypobaric physiology as part of their hypoxia awareness training. Before their practical session in the altitude chamber, the ascans are taught the signs and symptoms of hypoxia, the types of hypoxia, and situations which could lead to hypoxia, such as a rapid decompression of a spacecraft! After the theory lesson, ascans enter a hypobaric chamber (Figures 4.10 and 4.11) for a high-altitude indoctrination (HAI) flight (Panel 4.2) to 26,000 feet.

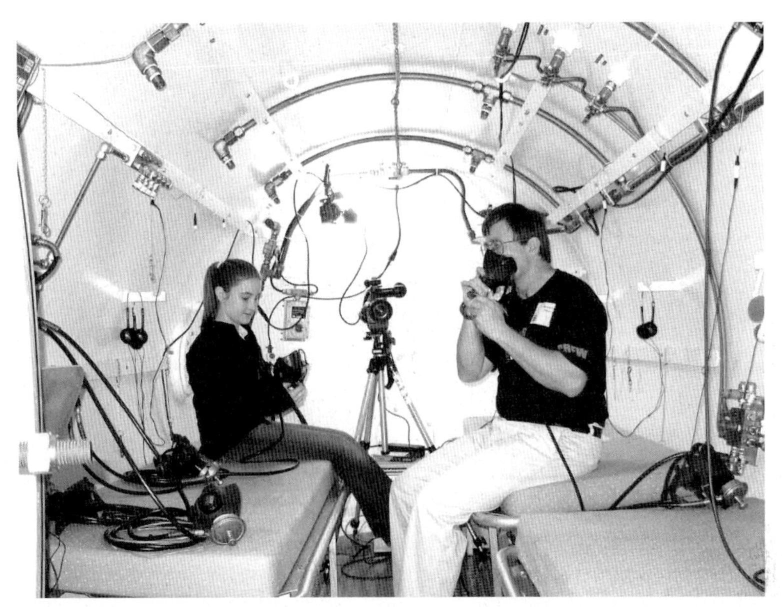

Figure 4.10. One of the checks in an ascan's probationary training is a ride in a hypobaric chamber. Here, Duncan Milne, Simon Fraser University's chamber engineer, tests the built-in breathing device with his daughter. The chamber was located in the author's lab. Image courtesy: Duncan Milne.

Figure 4.11. Simon Fraser University's hypobaric chamber. Image courtesy: Duncan Milne.

Panel 4.2. High-altitude indoctrination training

The HAI involves a 20-min prebreathe of 100% oxygen at sea-level, followed by a decrease in pressure to simulate an ascent to 2,450 m (8,000 feet) for equipment tests and ear checks. Following this, the chamber is brought to a simulated altitude of 7,950 m (26,000 feet) where the ascans remove their breathing masks and breathe ambient air (270 mmHg pressure) for up to 4 min. While at 7,950 m, the ascans perform arithmetic, cognitive, and hand coordination tasks. They also try to determine their initial symptoms of hypoxia. If they experience recognizable symptoms prior to the 4 min, they resume breathing from their oxygen masks. After 4 min, masks are donned and equipment checks are made, before descending to ground level.

The final training session before the first week of training is over is using the virtual reality (VR) trainer to learn how to execute a parachute landing fall before practicing for real. After a few bumpy landings, the ascans look forward to doing some flying, albeit in a turboprop.

T-34 ground school

Ascans spend the second week of training preparing for their first T-34 flights. Still based in Pensacola, the T-34 training, as with most astronaut training, is split between classroom and simulators. First, the ascans learn the systems of the aircraft before moving on to communication procedures, pre-flight, flight, and post-flight checklists, in addition to the emergency procedures. In between attending T-34 ground school, the ascans also spend time in the cockpit simulator. The T-34 theory week is an intense training increment, requiring the ascans to not only learn, but memorize several procedures before they can actually fly the aircraft.

T-34 flight training

After a week of ground school and simulator training, the ascans look forward to actually flying the T-34 (Figure 4.12). After reviewing planned flight and emergency procedures with their instructors, they conduct the pre-flight inspection, taxi on the apron, and finally take off. During the 4 weeks they spend learning how to fly the T-34, the ascans learn how to perform all the skills required by a civilian pilot studying for his private pilot's license.

After 6 weeks of training, the ascans return to Houston for a couple of weeks. During this period, most ascans are in the process of moving their families to Houston. The 2 weeks spent at Johnson Space Center (JSC) are mostly for

Figure 4.12. T-34 training aircraft. Image courtesy: NASA.

indoctrination to the world's most famous space agency. During their time at JSC, the ascans have the opportunity to meet NASA engineers working on ISS hardware and the Orion Program. They also have the opportunity to meet veteran astronauts. For example, the Class of 2004 was fortunate enough to meet astronaut legend, John Young, who gave the group a presentation on lunar exploration.

Survival training

Next for the ascans is a week-long survival training program at the Navy's wilderness site near Rangeley, Maine. The training begins early on a Monday morning with an introduction to the survival skills the ascans will be taught, followed by gear issue. The survival gear is standard military issue and includes standard camouflage clothing, two canteens, a bayonet, iodine tablets, a poncho with liner, a compass, and a map (Figure 4.13). After wrapping the items into a makeshift backpack made from a military issue coat, the ascans head off into the training area to learn how to survive. On arrival at Alpha Base, the ascans receive a military issue ready-to-eat meal and are shown a nearby stream from which to get water. Next is a classroom session for the inevitable "death by Powerpoint" presentation dealing with the standard survival skills for finding and trapping food, finding water, fire starting, shelter building, medical emergencies, and land

Figure 4.13. NASA's Class of 2004 ascans work together as they test their navigation skills in the wilderness of Maine during land survival training. Image courtesy: NASA.

navigation. Once the theory lessons are over, the ascans divide into groups of three or four and spend the first night in small wooden shelters covered by parachute fabric.

The training at Rangeley has two purposes. Astronauts fly a minimum of 100 hr a year in T-38s to maintain aviation and navigation proficiency. The flights usually take them over deserts and mountains. If they had to eject from the aircraft, they might have to fend for themselves for at least a few hours and possibly longer. The training also provides a unique and challenging teambuilding environment for a group that is together for the better part of a year. When they are assigned to spaceflight crews, they will form other close-knit teams. Given the varied backgrounds of the ascans in the class, learning leadership and team-working skills while dealing with adversity proves an invaluable experience.

During the week of instruction, the ascans receive briefings and instruction from a team of eight Naval specialists, who teach the class everything from how to catch squirrels to making spruce tree tea. The Navy instructors also emphasize the philosophy that ingenuity is the key to survival and explain the multiple uses of the limited materials and equipment the ascans have been given. For example, the poncho, while unnecessary in the absence of rain, can serve as a shelter, water collector, and even a sleeping bag. Similarly, parachute cord has almost as many uses as duct tape.

For 3 days, the instructors guide the group through survival medicine, signals, fire

Figure 4.14. NASA Class of 2004 ascans "rescue" a classmate during survival training. Image courtesy: NASA (*see colour section*).

craft, and navigation. As the training progresses, the ascans learn to search for evacuation sites and work with rationed materials to create a group shelter. At this stage, they can forage through the woods, and are capable of recognizing edible plants. They have also been briefed on edible insects and wildlife, but most prefer the military issue meal.

As with most phases of astronaut training, the survival training ends with an examination that begins at 5.30 in the morning on the last day. The exam is a staged rescue scenario, requiring the ascans to evacuate two of their colleagues to a rescue location, using improvised litters and ponchos (Figure 4.14).

Flying the T-38

Next on the ascans' agenda is a trip to Ellington Field, in Houston, for a week of ground school on NASA's T-38 training jet (Figure 4.15). The T-38 is a twin-engine jet trainer that has been used in the United States Air Force (USAF) since 1961 to train fighter pilots. Flight training is conducted with a pilot flying the jet trainer, as a mission specialist in the rear seat communicates with air traffic control and performs navigation.

Apart from being a fun jet to fly, trained astronauts use the T-38 jet to travel

Figure 4.15. Two T-38s fly over Edwards Air Force Base. Image courtesy: NASA (Jim Ross).

between NASA facilities to save time; it's a bit like having a company car! The comparison between flying the T-34 and the T-38 is like night and day. Whereas the T-34 is a turboprop with a maximum speed of a little over 400 km/hr, the T-38 reaches its maximum speed of Mach 1.08 within 1 min of take-off and can climb to a ceiling of more than 9,000 m. Another difference ascans immediately notice is the acceleration. Whereas the T-34 tops out at 2.5 Gs, the T-38 is capable of exerting a whopping 5 Gs on its two-person crew, making the jet useful for training astronauts for the intense G-forces encountered during a mission. Another difference is the means of egress; in the T-34, ascans have to climb out onto the wing and jump off the back, but in the T-38, egress is by means of a rocket-loaded ejection seat!

For most ascans, their first flight in the T-38 is the first time they have traveled supersonic. Usually, for fun, the instructors throw in a zero-G maneuver, allowing ascans their first taste of what spaceflight will feel like. After their initiation flight, the ascans move on to more essential training such as crew resource management (Panel 4.3), a skill they will use when they finally fly in space.

NASA Center acquaintance visits

After experiencing the thrills of the T-38, ascans spend time becoming acquainted

Panel 4.3. Crew resource management

CRM training originated from a NASA workshop in 1979 that focused on improving air safety. The NASA research found that the primary cause of most aviation accidents was human error, and the main problems were failures in interpersonal communication, leadership, and decision making in the cockpit.

CRM training encompasses a wide range of knowledge, skills, and attitudes, including communications, situational awareness, problem solving, decision making, and teamwork. It is basically a management system that makes optimum use of all available resources, including equipment, procedures, and people, to ensure the safety efficiency of flight operations.

From a space mission perspective, CRM is concerned not so much with the technical knowledge and skills required to fly and operate a spacecraft, but rather with the cognitive and interpersonal skills needed to manage the flight within an organized spacecraft system. In this context, cognitive skills are defined as the mental processes used to gain and maintain situational awareness, to solve problems, and to make decisions.

with some of NASA's field centers. Their first stop on the center tour is usually NASA Headquarters, located in Washington, DC. After visiting NASA HQ, the ascans move on to Goddard Flight Center (GFC) in Maryland, where they learn how NASA develops and operates unmanned scientific spacecraft. From Goddard, the ascans fly to NASA's Glenn Research Center (GRC) in Cleveland, Ohio, where they have an opportunity to talk to scientists developing new propulsion systems and researchers studying zero-gravity.

Zero-gravity

After the whistle-stop center tour, the next training phase is microgravity indoctrination. To simulate microgravity (Panel 4.4), NASA flies a specially modified aircraft out over the Gulf of Mexico, where it performs a series of parabolas (Figure 4.16). For many of the ascans, parabolic flight is one of their favorite training experiences.

Familiarization training

After parabolic flight training, ascans will usually spend more time becoming acquainted with other NASA Centers. One center they will almost certainly visit is the Stennis Space Center (SSC) in Mississippi, where engines for NASA's new launch

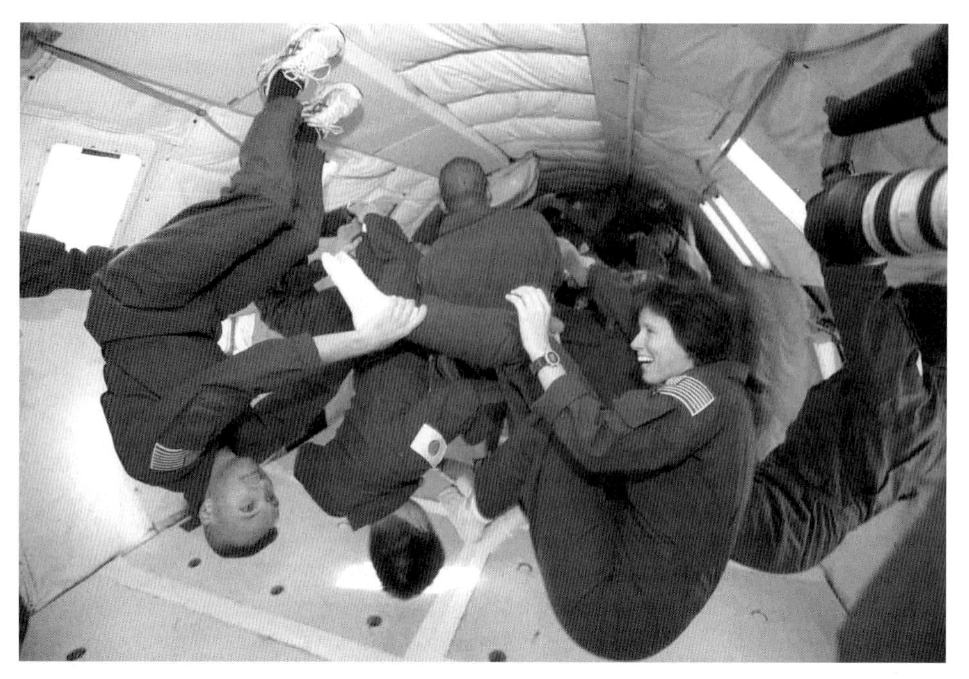

Figure 4.16. Some of NASA's 2004 class of astronaut candidates and some JAXA astronauts tumble during one of a series of reduced gravity (zero-G) sessions. Image courtesy: NASA.

Panel 4.4. Zero-G

Microgravity can be simulated when an aircraft flies a Keplerian trajectory, or parabolic flight profile that requires the aircraft to climb rapidly at a 45° angle and then follow a path called a parabola. At the apex of the parabola, the aircraft produces a near-zero-G effect (1×10^{-3} G) for between 20 and 28 sec, just as the aircraft achieves 9,500 of the 10,000-m ascent (termed *pull-up*) before it slows. The aircraft then traces a parabola (pushover), descending rapidly at a 30–45° angle (termed *push-out*) to 7,300 m.

The acceleration forces produce approximately two times normal gravity (2-G) during the "pull-up" and "pull-out" phases of the flight, and between these phases is an intermediate phase, termed the "pushover" that occurs at the apogee, generating a zero-G environment with less than 1% of Earth's gravity. Due to the gut-wrenching sensations produced during parabolic training flights, the aircraft has earned the nickname "vomit comet".

vehicles are developed and tested. Another center on the agenda is usually the Marshall Space Flight Center (MSFC), located in Huntsville, Alabama. Here, they

Figure 4.17. Artist's rendering of *Ares I* on the launch pad. Image courtesy: NASA.

learn how MSFC is developing NASA's next-generation launch vehicles, *Ares I* (Figure 4.17) and *Ares V*, which will transport crew, supplies, equipment, water, fuel, and hardware, initially to the ISS, and later to the Moon. The ascans also learn how MSFC plays a significant role in the ISS Program, by developing systems to support life on the station, managing logistic modules for transporting science experiments, and managing the Payload Operations Center. Of particular interest to the new generation of space explorers will be NASA's new Altair lunar lander, which will eventually return astronauts to the lunar surface.

Orion and Ares I systems

On returning to Houston, ascans begin a lengthy period of training that focuses on the systems of the *Ares I* launch vehicle and *Orion*, the crew exploration vehicle. At this stage, the ascans have completed almost 9 months of training. In what is mostly classroom study, the ascans learn about the major systems (Table 4.2) from instructors who are mostly engineers from NASA's Training Division.

Table 4.2. Major systems training.

Main Propulsion System	Electrical Power System
Data Processing System	Auxiliary Power Unit/Hydraulic System
Orbital Maneuvering System	Communication System
Reaction Control System	Caution and Warning System

The objective of the training is to ensure each ascan is capable of becoming a proficient operator, whether they are a pilot, mission specialist, or a commander. To prepare for the long 4-hr classes, ascans are issued a stack of textbooks and technical manuals more than a meter high! Just like a university course, the ascans have to take tests at the end of each stage. The first test has about 200 questions, divided into 14 sections for the various systems. It's a long test, taking 2 hr to complete. Any ascan scoring less than 80% on any section is required to complete remedial training.

Once they have successfully completed the first major systems phase, the ascans move on to a more practical phase in which they learn how to actually operate the equipment. This phase of training is completed by having the ascans utilize high-fidelity mockups called Single Systems Trainers (SSTs). The SST is a mockup of *Orion*'s flight deck, with all the computers, switches, and flat-screen displays. As the name implies, the SST allows the ascans to work one system at a time. As with their first major systems training, this phase of training finishes with testing. In addition to a written exam, the ascans must also complete a performance-based test that takes place inside the *Orion* SST. This test confirms to the instructors the ascans can apply the book knowledge to the real world of spaceflight.

Not surprisingly, studying for all these examinations is an arduous task, somewhat akin to cramming for a Medical College Admission Test (MCAT). As with prospective medical students who must learn by rote thousands of items of information, study techniques vary. Some ascans study in small groups, others study alone, while others make hundreds of flashcards.

Increasing the stress on the ascans during their major systems instruction is the requirement to maintain their T-38 flight hours. Every year, each ascan must log

100 hr of flying in the T-38. While this equates to only 8 or 9 hr a month, which doesn't sound like much, in reality, it is quite a challenge for busy ascans to stay current. First, the pilot and mission specialist must meet up 1 hr before the flight to plan the flight and review safety procedures, before actually conducting the flight. Once the time required to drive to the airport and back is factored in, the actual time required to log just 1 hr of flying easily approaches 3 hr!

Acquaintance visit

Once the studying of major systems is over, ascans head out for another acquaintance visit, this time to White Sands Test Facility (WSTF) in New Mexico. WSTF is home to the Tracking and Data Relay Satellite system (TDRS), which enables astronauts to communicate with Mission Control. Next on the acquaintance schedule is a visit to Langley Research Center (LRC) in Hampton, Virginia, where ascans have the opportunity to meet scientists, such as Dr. Joel Levine, who are developing exploration vehicles for Mars. The ascans also visit LRC's Mach 10 wind tunnel, where scale models of the *Ares I* and *Ares V* are tested. In the Data Visualization and Analysis Lab, ascans are provided with an overview of the research being conducted in the field of 3D modeling, which is helping scientists understand the effects of radiation.

After their visit to LRC, the ascans travel to Kennedy Space Center (KSC), where they visit Complex 39 B, from where, eventually, an *Ares I* will launch them into space. They also take a tour of the Vehicle Assembly Building (VAB), where the Shuttle used to be stacked, but which has since been modified for NASA's new family of launch vehicles. Also on the tour is the Space Station Processing Facility (SSPF), where ISS hardware is stored and tested before being flown.

Spacesuit indoctrination

After 9 months of astronaut training, the ascans have yet to slip on a spacesuit but, after the latest round of acquaintance visits, they finally have the opportunity to be custom-fitted for one of NASA's famous pumpkin suits (Figure 4.18).

The Advanced Crew Escape Suit (ACES) is a full-pressure suit designed to protect astronauts in the event of loss of cabin pressure at altitudes up to 30 km. It also insulates the crew from cold air or water temperatures, in the event of a bailout. Actually donning the suit is one of the biggest challenges an astronaut faces on launch day, so ascans take every opportunity in training to become familiar with all the clothing garments. Assisted by a suit technician, the ascans first put on a diaper (Figure 4.19), or a Maximum Absorbency Garment (MAG – yes, NASA has an acronym for *everything*!), to use the correct NASA term!

Next, they pull on a layer of lightweight polypropylene underwear and some long-johns fitted with plastic tubes woven into the garments, which circulate water to keep crewmembers cool in what is a very warm suit. After the long-johns comes the G-

Figure 4.18. Astronauts sit in the suit prep room in the Advanced Crew Escape System (ACES) suits. Image courtesy: NASA (*see colour section*).

Figure 4.19. The Maximum Absorbency Garment: the one piece of flight clothing astronauts would rather not talk about! Image courtesy: NASA.

suit, a one-piece pressure garment assembly with integrated pressure bladders and ventilation system. Then the helmet is screwed onto the neck-ring, and ascans check their oxygen connector on their left thigh is connected via a special connector at the base of the neck-ring. The full-pressure helmet comprises a locking clear visor and a black sunshade worn to reduce any glare from reflected sunlight, especially during the approach and landing phases of the mission. A dark-brown communications cap is worn underneath the helmet, and connected to a special plug inside the helmet, which is then connected to the intercom system in *Orion*. An anti-suffocation valve at the back of the helmet allows for the passing of carbon dioxide from the helmet. Once the anti-suffocation valve has been checked by the suit technician, the helmet's pressure face-plate is locked into place, using a mechanical seal with a prominent "lockdown" bar. In common with the helmet, the suit's gloves are also attached using a locking ring in matching orange. To enable crewmembers to flick switches and turn knobs, the palms of the gloves are textured. Once the gloves are fitted, the ascans slip on a pair of heavy black paratrooper-style boots with zippers instead of laces. You might think the ascan is fully clothed in their suit by now but you'd be wrong. After the boots, the ascan still has to don a survival backpack, which includes a personal life raft. At the end of what is by all accounts quite a workout, the ascan is wearing a suit weighing 35 kg!

Я не говорю по-русски ***[ya nye go-vo-RYOO pah ROO-ski]****

Ten months after they began their training, the ascans take a 2-week introductory program in Russian language and culture. The first class is an introduction to the seemingly indecipherable Cyrillic alphabet, which has many ascans wishing they were back in survival school!

Those unfamiliar with the ISS program may be wondering why American ascans would need to learn Russian. The answer is relatively simple. Russia is one of NASA's many ISS partners. Most significantly, Russia is the only partner with a manned spaceflight capability. Following the *Columbia* tragedy, NASA relied on Russia to ferry American and partner astronauts to the ISS. Also, once the Shuttle is retired in September, 2010, NASA will once again rely on the aging Soyuz to transport astronauts into LEO, at a cost of $51 million per astronaut! Because of this partnership, American astronauts spend much of their time traveling to Star City, in Russia, for familiarization training on Soyuz systems.

Scuba-diving

Following basic Russian language training, the ascans move on to their scuba-diving qualification. In common with so many phases of training, scuba-diving utilizes one

---

* I don't speak Russian in Russian!

Figure 4.20. Equipped with scuba gear, NASA's 2004 class of astronaut candidates participate in a training session in the Neutral Buoyancy Laboratory. Image courtesy: NASA (*see colour section*).

of NASA's many simulators. The Neutral Buoyancy Laboratory (NBL) housed within the Sonny Carter Training Facility (SCTF) at JSC is 61 m long and 12 m deep, and holds 23.5 million liters of water. In addition to training ascans in scuba-diving techniques (Figure 4.20), the NBL is also used to train astronauts for EVAs.

Scuba-diving is one of the last practical training increments. With 8 months left until completion of their initial training, the ascans spend more and more time in classrooms studying all aspects of *Ares I* and *Orion* vehicle operations in greater detail than was covered during the initial major systems training phase.

In between studying technical manuals, ascans have the opportunity to train in the *Orion* motion-based simulator, located at the Jake Garn Training Facility (JGTF) at JSC. The motion-based trainer simulates the vibrations, noise, and views the ascans will experience during an *Ares I* launch and landing, while a fixed-base simulator is used for rendezvous and payload operations training. JGTF also houses a functional space station simulator, which familiarizes the ascans with the laboratory systems of the ISS.

Another simulator the ascans use during these final few months of training is the *Orion* vehicle mockup (Figure 4.21). In common with the JGTF, the *Orion* mockup at JSC consists of components that prepare astronauts to fly and operate the crew exploration vehicle (Panel 4.5).

The mockup includes a flight deck complete with high-fidelity components, such as the Honeywell flat-screens, panels, seats, and lights. A third mockup ascans become familiar with is the Space Station Mockup and Training Facility (SSMTF), a

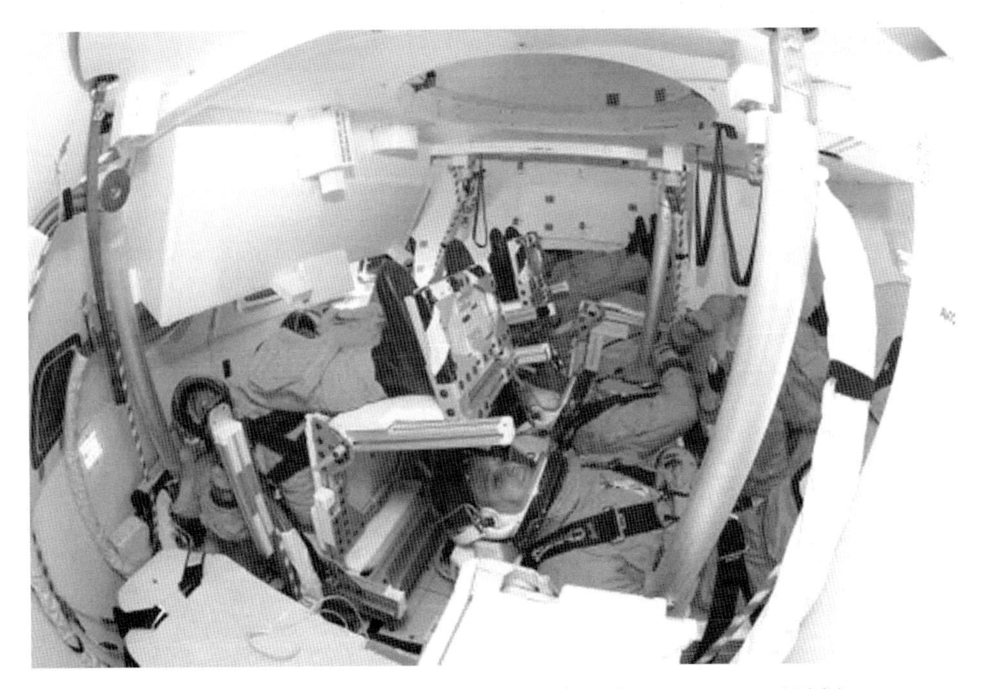

Figure 4.21. Astronauts test an *Orion* mockup. Image courtesy: NASA.

full-scale replica of the ISS, providing as much realism as possible to match conditions ascans will eventually experience onboard the orbiting space station.

Panel 4.5. *Orion*'s cockpit

Orion features a start-of-the-art flight-control system consisting of three briefcase-sized Honeywell Flight Control Modules. Two of the control computers could completely fail and the third would still be able to fly the vehicle. In the worst-case scenario of a complete power failure, *Orion* carries an emergency system powered by independent batteries, providing the crew with enough capability to bring the vehicle back safely. In common with fighter aircraft cockpits, *Orion*'s control systems rely heavily on *sensor fusion*, a type of automation, relieving the astronaut pilot of being a sensor integrator and allowing him/her to focus on the mission. Such a system makes sense, given that many astronaut pilots came from advanced cockpits such as the F-15 and F-22.

In *Orion*'s cockpit, astronaut pilots will be able to change displays as if on a revolving panel, thanks to four flat-screen displays, each about the size of a large desktop monitor. During ascent to orbit, the displays will operate similarly to the screens in conventional airliners. One display will show an artificial horizon, another will display velocity, and a third will show altitude. The fourth display will show life support status and communications information. Once *Orion* reaches orbit, the displays will change to readouts, showing rendezvous and docking information such as the vehicle's flight path, range, and rate of closing to the ISS.

Finally, after approximately 18 months of training, ascans become astronauts, with the presentation of silver wings. Although the wings signify the end of what sometimes felt like a series of endless training cycles, astronauts must still wait a long time before they actually fly into space and trade the silver wings for gold ones.* Before the realization of the ultimate dream, they must first enter the world of technical assignments and public relations activities and wait for that coveted flight assignment.

* Whereas the silver wings are bought out of the coffee fund, astronauts wanting to buy real gold wings must pay about $400.

5

Technical assignments

Someone has to be first, and someone has to be last. That is the reality of mission assignments. The question many astronauts ask is how to get to the front of the line. One way is to excel in their technical assignments. Typically, newly minted astronauts, in their search for the coveted flight assignment, tackle their technical assignments energetically. They also redouble their efforts academically, plowing their way through reams and reams of technical data in an effort to impress those making the mission assignment decisions. In between conducting their technical assignment duties and wading through weighty technical tomes, the astronauts take every opportunity to display teamwork and initiative to those in the Astronaut Office, where the powers that be make the flight assignment decisions.

TYPES OF TECHNICAL ASSIGNMENTS

During the course of their career, an astronaut will hold several technical assignments, usually between missions and mission training. These assignments may include such wordy positions as Technical Assistant to the Director of Flight Crew Operations to Astronaut Office Liaison to the Safety, Reliability and Quality Assurance Directorate. NASA, and space agencies generally, rarely have short job titles! Since it is impossible to describe every type of technical assignment, this chapter restricts itself to some of the more high-profile jobs an astronaut can hold while waiting for their flight assignment.

Space Operations Mission Directorate

Many technical assignments are within the Space Operations Mission Directorate (SOMD). The SOMD provides NASA with leadership and management of NASA space operations related to human exploration in and beyond low-Earth orbit (LEO). The directorate (Figure 5.1) also oversees low-level requirements development, policy, and programmatic oversight in addition to supporting both human and

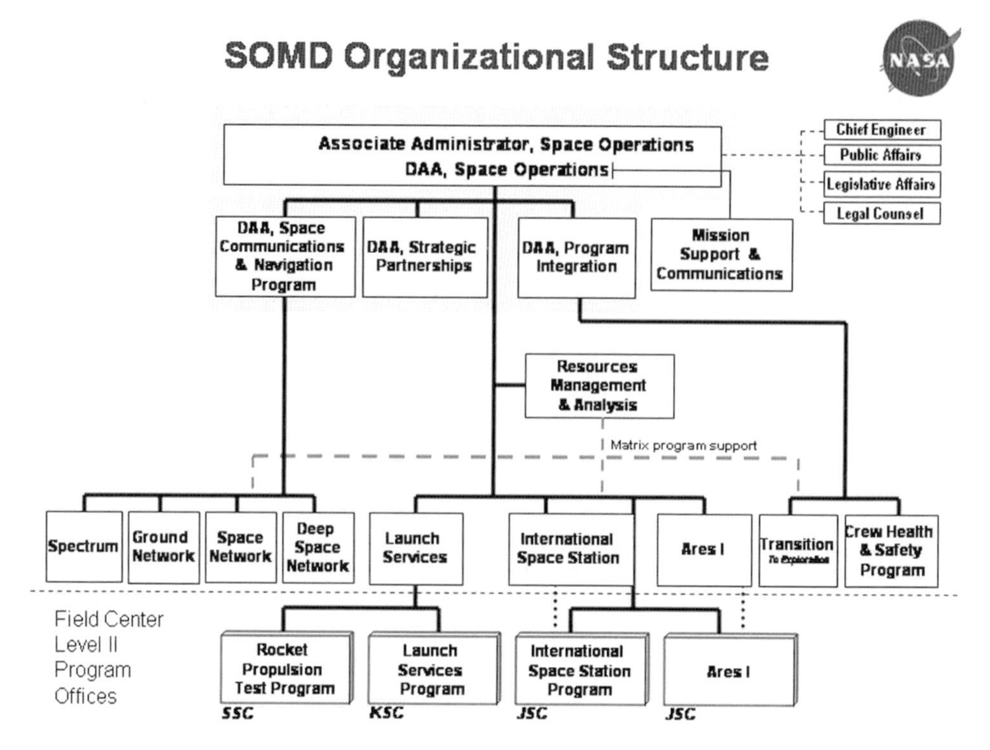

Figure 5.1. Space Operations Mission Directorate organizational structure. Image courtesy: NASA.

robotic exploration programs. Astronauts assigned to SOMD may act as either a special or technical assistant, depending on which program they are assigned to. Given the myriad responsibilities of SOMD, astronauts assigned to the directorate could be tasked with anything from the International Space Station (ISS) Payloads Office to the Extravehicular Activity (EVA) Project Office. An assignment to any SOMD department is a useful experience for a neophyte astronaut, as it not only provides an insight into the operational capabilities of the space operations program, but also provides an introduction to some of the many non-NASA participants involved in the ISS Program.

International Space Station Payloads Office

Being assigned to the ISS Payloads Office isn't such a bad assignment, since the office is responsible for the development and delivery of the payload rack facilities the astronauts will use on orbit. Because the office is responsible for so many payloads, an astronaut may be assigned to manage a particular rack such as the Human Research Facility (HRF).

Human Research Facility

The HRF provides an on-orbit laboratory enabling life science researchers to study and evaluate the physiological, behavioral, and chemical changes in human beings induced by space flight. Research performed with the HRF will provide data relevant to longer-term adaptation to the space flight environment. Astronauts assigned to facilities such as the HRF will usually serve as a technical assistant. In this capacity, the astronaut may be charged with the management and dissemination of data when they are returned to Earth and deciding which experiments to fly on the next mission. In addition to dealing with the investigative aspects of the facility, the astronaut will also be responsible for ensuring any deficiencies are dealt with on the ground and that parts are manifested on the next flight. This task could involve anything from ensuring the ISS crew has sufficient calibration gases for the HRF's Analyzer Module to checking each crewmember has enough saliva filters for all the experiments.

European Physiology Module

Another science facility an astronaut may be involved with is the European Space Agency's (ESA) European Physiology Module (EPM), located in the *Columbus* module. The EPM comprises a set of experiments used to investigate the effects of long-duration spaceflight on the human body. Experiment results will eventually contribute to an increased understanding of age-related bone loss, balance disorders, and other ailments back on Earth. Since the EPM is designed to accommodate dozens of experiments during the lifetime of the ISS, new science equipment must be manifested onboard the Soyuz spacecraft in accordance with research being conducted onboard. Also, since many of the samples can only be analyzed back on Earth, these must also be manifested. Another of the astronaut's chores may be to ensure data collected prior to, during, and post flight are made available to researchers on the ground.

Payload integration

Another assignment within the ISS Payloads Office may be working with the payload integration team that manages and prepares payloads bound for the ISS. The payload integration team is responsible for helping payload developers prepare their payloads and ensuring items get to the station in one piece. Although the job might sound relatively simple, as with many aspects of the manned space program, the process is actually quite complex.

One of the reasons for payload integration being such a difficult undertaking is that space station payloads come from a variety of sources. Many are developed internally by organizations within NASA while others are the result of proposals from industry and academia. These payloads are developed by researchers who have had proposals selected to fly experiments in space. NASA also carries payloads to the station for the other nations involved in the ISS Program.

Payloads may include everything from scientific experiments to new station hardware to equipment that will be used by the crew. Part of the job of an astronaut assigned to payloads may be to work with items that fly not only on American launch vehicles, but ones that fly on other spacecraft, such as the Russian Progress cargo vehicle. They may also work in the complex process of certifying a payload for flight. For example, a relatively simple payload such as a box of seeds requires a biohazard safety review to be processed. Certification also means dealing with, and complying with, dozens of policies and requirements, such as those listed in Table 5.1.

Table 5.1. Examples of payload policies and requirements.

Policy	*Description*
NSTS 1700B	Safety Policy and Requirements for Payloads Using the Space Transportation System and the International Space Station
SSP 52005C	Safety Critical Structures Requirements and Guidelines
NSTS 14046E	Payloads Verification Requirements
ECSS-Q-70-36A	Space Product Assurance – Material Selection for Controlling Stress Corrosion Cracking
MSFC-STD-3029	Guidelines for the Selection of Metallic Materials for SCC Resistance
P32928-103-2001	Requirements for International Partners Cargoes transported on Russian Progress and Soyuz Vehicles

First, the astronaut must verify the payload. This requires assessing the payload's structural elements, including associated interfaces, fasteners, and welds in the payload. Next, the components' primary load path, including pressure systems, uncontained glass, rotating machinery, mechanical stops, and containment devices, must be assessed. In payload parlance, the primary load path is the collection of structural elements that transfer load from one part of a structure to another. Elements in the primary load path experience loading in excess of that created by their own mass. Once the verification steps are completed, a Structural Verification Plan (SVP) is generated. The SVP includes detailed definition of the structural design, the loads to be considered for the verification, and all the testing the different hardware will undertake. The SVP must then be passed on to NASA's Structures Working Group (SWG), which means more meetings and more paperwork.

For astronauts who have just completed their stint as ascans, being assigned to payloads can be a little demoralizing, since it usually means spending a year or more in offices conducting meetings. While their colleagues excitedly trade stories about spacewalk procedures and developing remote manipulator protocols, the astronauts assigned to payloads pray they won't be asked how their days in the briefing rooms went!

After determining the structural elements and the load path, attention is turned to the payload design. This is a laborious process involving poring over data consisting of drawings, parts, materials, and processes information that can be used to manufacture and assemble a payload. The next step is to verify the payload, which involves myriad analytical and testing activities performed to verify the adequacy of the payload with regard to safety and mission success. The process of verification inevitably means complying with all sorts of policies, requirements, and procedures. After the verification process, attention is turned to assessing the design limit loads, after which the document is submitted for review and the payload developer provides a detailed model of the payload, which is used for analytical verification. The astronaut working in payloads may be tasked with making sure that the model is checked for mathematical correctness and ensuring it is validated by means of a series of dynamic and/or static tests.

The astronaut might think they're nearing the end of the road at this stage, but just the testing can take months. Once the myriad tests are conducted, the payload enters the assembly and installation phase, which requires more testing, ground handling, and transportation. This phase also involves assessing the effects of lift-off, descent, re-entry, landing, and emergency landing scenarios and consideration of factors such as the acoustic and thermal environments and other constraints such as atmospheric pressure. Since there are so many people involved in the integration process, the payloads astronaut spends most of their time shuttling from one meeting to the next. A typical day might start at 7.30 in the morning and finish after 6.00 in the evening. Due to the overwhelming amount of paperwork, astronauts inevitably have to take work home, not only in the evenings, but during the weekend as well. Compounding the amount of work is the unfamiliarity of the material. For an astronaut who used to fly fighter planes or has a Ph.D. in astrophysics, the labyrinthine bureaucracy required to certify and verify a payload is about as alien as the surface of Mars. But, because they're astronauts, they're expected to become experts quickly, which means more and more homework. One morning they may have to present a talk on the merits of metallic materials with high resistance to stress corrosion, while in the afternoon they may be expected to be the subject matter expert on composite end-of-life properties and creep behavior! After talking about creep behavior (which happens to be the tendency of a solid material to slowly move or deform permanently under the influence of stresses!), the astronaut may then have to take home volumes of references describing such strange objects as the peculiarities of static and dynamic envelopes and payload interface requirements.

Once the payload is finally ready to fly, pre-mission planning and coordination are necessary to define the requirements for ground support. This phase includes selecting a number of services required once the payload is actually on the ISS. For example, the payload may require ground-to-space voice loops and ISS downlink video capabilities.

While a technical assignment to payloads might not sound the most exciting job for a newly trained astronaut, they do have an opportunity to work directly with the astronaut onboard the ISS and provide them with information they need when they are ready to use the payloads. However, many astronauts assigned to payloads, while recognizing the value of the experience, are happy to move to an assignment that offers a little more excitement, such as the EVA Project Office.

Extravehicular Activity Project Office

Perhaps one of the most coveted technical assignments is working in NASA's EVA Office. EVA includes activities from the ISS as well as future lunar and Mars surface operations. From the earliest preparations on the ground, until astronauts are actually working in the vacuum of space, a lot of effort goes into a spacewalk. Since most astronauts out of training will eventually conduct EVAs, seeing how spacewalk preparation is planned provides them with a useful insight into the activities they will be eventually performing onboard the ISS. Part of the job may be developing procedures allowing astronauts to accomplish tasks for a specific mission, while other duties may involve refining contingency tasks, reducing the risk of decompression sickness (DCS), or testing new spacesuits such as the Mark III (Figure 5.2).

Figure 5.2. Spacesuit engineer, Dustin Gohmert, simulates work in a mock crater in JSC's Lunar Yard. His mode of transport, NASA's prototype lunar rover, stands in the background. Image courtesy: NASA.

The Mark III suit is a NASA spacesuit technology demonstrator built by ILC Dover. While heavier than other suits (it weighs 59 kg!), the Mark III is very mobile thanks to a mix of hard and soft suit components, including hard upper torso, hard lower torso, and hip elements made of graphite/epoxy composite, and soft fabric joints at the elbow, knee, and ankle. Thanks to a 57-kPa operating pressure, the Mark III is a zero-prebreathe suit. This means astronauts transition from a one atmosphere, mixed-gas space station environment, such as on the ISS, to the suit, without risk of the bends, which can occur with rapid depressurization from an atmosphere containing nitrogen or another inert gas. Currently, astronauts must spend several hours in a reduced pressure, pure oxygen environment before EVA to avoid these risks. Recently, the Mark III has been involved in field testing during NASA's annual Desert Research and Technology Studies (D-RATS) field trials, during which suit occupants (astronauts and engineers) interacted with one another, and with rovers and other equipment.

CapCom

One of the most popular technical assignments among astronauts is working in the Mission Control Center (MCC) as a CapCom (a name derived from the term 'CAPsule COMmunicators' during NASA's earliest missions in space, when spacecraft were known as capsules). Amidst the hundreds of people involved in coordinating every aspect of a mission and its crew, not everyone can speak to the astronauts in space. This would only lead to confusion and costly, even fatal, errors. To ensure efficient communication, NASA uses CapComs, who act as the orbiting astronauts' sole voice from Mission Control.

The ISS MCC supports the myriad activities onboard the station and ensures a coordinated, consistent, and efficient line of communication. This is achieved by employing a team of CapComs (Figure 5.3) who are on duty 7 days a week. Since the ground-to-space voice communication link is a limited and precious resource, CapComs must first and foremost be excellent communicators. Voice communications must be concise and clear, while, at the same time, provide all the necessary information the onboard crew needs to correctly perform their duties and understand any given situation. CapComs have a broad understanding of all ISS systems, enabling them to participate in discussions held in the MCC and troubleshooting planning. As a rule, flown astronauts work as CapComs, since they have an understanding of what the flight crew is going through at any given time. Their training and experience give them the ability to ensure directions given to the on-orbit crew are practical and consider human factors. However, there have been several astronauts who have held the CapCom technical assignment without having flown. One example was former astronaut, Donald Thomas, who served as CapCom for Shuttle missions STS-47, 52, and 53, before flying on STS-65 in July, 1994.

Figure 5.3. NASA's Mission Control Center. Note the position of CapCom on the right. Image courtesy: NASA.

Flight Crew Operations Directorate

The Flight Crew Operations Directorate (FCOD) is responsible for overall planning, direction, and management of flight crew operations and various program activities at Johnson Space Center (JSC). These responsibilities include, but are not limited to, selecting and training astronaut candidates, determining flight crew training and flight crew simulation requirements, recommending specific flight crew assignments and training, and certifying payload specialists. Since the FCOD plays a role in mission assignments, astronauts assigned here do their best to make a good impression, in the hope their name might move up the list of names waiting for a flight. In addition to trying to get noticed, astronauts may also participate in the development of timelines and procedures, provide flight crew perspective in the development of new programs, and contribute to the development, acquisition, maintenance, and safe operation of astronaut training.

Office of the International Space Station Program Scientist

Another useful technical assignment for newly qualified astronauts is working in the office of the ISS Program Scientist, who is usually a former astronaut (e.g. Donald

Thomas became the ISS Program Scientist shortly after retiring as an astronaut in 2003). One of the most important jobs of an astronaut working onboard the ISS is to conduct experiments, so an assignment to work in an environment where investigations are developed is an invaluable experience. Although the amount of science was restricted while only three crewmembers occupied the ISS, the pace of scientific investigations will dramatically increase now the station has a permanent crew of six onboard. Also, thanks to the station's laboratory and research facilities being tripled in 2008, with the addition of ESA's *Columbus* and JAXA's *Kibo* scientific modules joining NASA's Destiny laboratory, the facilities available will considerably improve the scope of investigations in which astronauts will be involved in the future.

Astronauts assigned to the ISS Program Scientist are involved in all aspects of the myriad and convoluted processes leading to approving a science experiment for flight. In terms of paperwork and meetings, the process isn't much different from the route taken by the payload integration team. For example, astronauts assisting the ISS Program Scientist must deal with aspects as diverse as science payload documentation to experiment objectives and from inflight procedures to experiment hazard evaluation analysis. Since missions to the ISS are for 6 months, each astronaut is trained in conducting a number of science investigations. Some of these require routine data collection, others require the astronaut to participate as a research subject, while others just need the astronaut to monitor equipment from time to time. For example, the ISS Acoustics Measurement Program (AMP) requires astronauts to use sound-level meters to collect acoustics data from the ISS environment for a 2-hr period before downlinking the data to ground for analysis. The AMP experiment is responsible for ensuring a safe, healthy, and habitable acoustic environment on the ISS, in which crews can live, communicate, and work. This means ensuring the ISS is not too noisy, does not have irritating audible sounds, and does not have startling bursts of acoustic energy. In contrast, the Sleep–Wake Actigraphy and Light Exposure during Spaceflight experiments require astronauts to log their sleep using Actiwatches, which automatically collect data of sleep patterns. The Actiwatch is compared to sleep logs to investigate the affects of spaceflight and ambient light exposure on the sleep–wake cycles of the crew members during long-duration ISS stays. Another experiment requiring astronaut participation is the Mental Representation of Spatial Cues During Space Flight investigation, which assesses the effects of exposure to microgravity on the mental representation of spatial cues by astronauts during and after space flight. This investigation requires the astronauts to measure their depth and distance perception by means of geometric illusions, 3D scenes, hand-writing, and drawing tests. A more complex experiment is the Effect of Gravitational Context on EEG Dynamics study that tests prefrontal brain functions and spatial cognition in addition to the effect of gravitational context on brain processing. During this experiment, visual orientation and visuomotor tracking tasks, together with standardized electroencephalogram (EEG) tasks, are performed as a means of assessing general effects of the ISS environment on EEG signals.

Ensuring astronauts onboard the ISS are able to conduct all their science experiments means much work must be done on the ground, since the science must be accommodated within a hectic work schedule. To ensure the science for one particular investigation doesn't take too much of an astronaut's time, it is necessary to work closely with the principal investigators to ensure scientific and engineering requirements are clearly communicated. The Program Scientist's Office must also ensure the astronauts on ISS conduct the science as planned and that scientists are satisfied with the operations and results of their experiments.

Exploration Development Laboratory

Another popular technical posting is being attached to Lockheed Martin's Exploration Development Laboratory (EDL), where astronauts have the opportunity to test hardware they may be using to and from the Moon. Lockheed's EDL is a facility dedicated to supporting NASA's Constellation Program and, in particular, *Orion*, its new crew exploration vehicle. *Orion* is America's next-generation human spaceflight vehicle that will transport up to six astronauts to and from the ISS and up to four to the Moon and destinations beyond, beginning in 2015, following the retirement of the Space Shuttle. The laboratory is a state-of-the-art test facility funded by Lockheed Martin and its teammates, United Space Alliance (USA) and Honeywell. Located in Houston, adjacent to NASA's JSC, the EDL enables the Lockheed Martin team to take full advantage of early involvement and collaboration with astronaut flight crew members. By having astronauts test systems inside full-scale low-fidelity mockups (Figure 5.4), NASA and Lockheed Martin not only gain clarity on requirements, but also allow future crewmembers to work closely with engineers on issues ranging from emergency ingress and egress to vehicle fit, form, and function.

While working with Lockheed Martin's engineers, astronauts help test avionics and software, develop crew interfaces, and evaluate systems such as the Automated Rendezvous and Docking (AR&D) system. Astronauts also play an important role in the human factors design of *Orion*'s cockpit, such as evaluating panel displays (Figure 5.5), internal lighting seat location, crew stowage, hand controller positioning, and other human interface devices.

Russian Liaison

Given the amount of time astronauts spend away from their families every year, the technical assignment of Russian Liaison isn't the most popular assignment, since it means spending lots of time in Star City, near Moscow. Until recently, the Russian Liaison participated in the testing and integration of Russian hardware and software products developed for the ISS, which often meant working with the Energia Aerospace Company in Moscow. Russian Liaisons also spent a lot of time in Russia developing and verifying dual-language procedures for ISS crews in addition to overseeing inventory and loading of ISS resupply capsules. With the retirement of

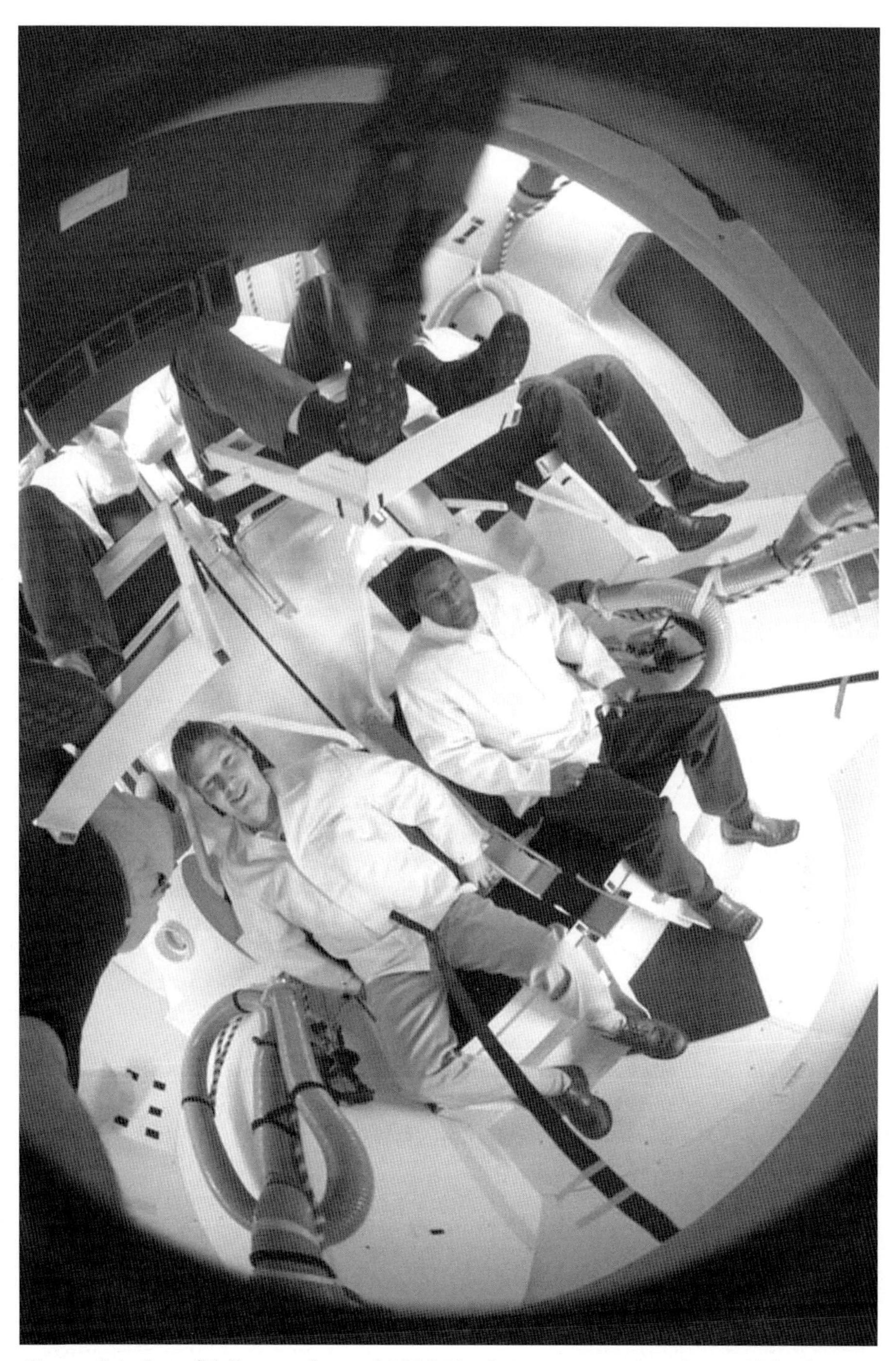

Figure 5.4. Low-fidelity mockup of NASA's *Orion* crew exploration vehicle. Image courtesy: NASA.

Figure 5.5. Panel display inside *Orion* mockup. Image courtesy: NASA.

the Shuttle in 2010, the role of NASA's Russian Liaison astronauts will take on even greater significance, since NASA astronauts will be hitching rides to the ISS onboard the Russian Soyuz spacecraft until at least 2015.

Perhaps one of the greatest challenges of future Russian Liaison astronauts will be developing dual-language procedures. Although NASA astronauts would welcome English as the single language of operation, this isn't going to happen as long as the Russians are providing the ride into space. Also, in the interests of safety and operation, it is crucial both astronauts and cosmonauts have a clear understanding of all procedures and plans, which is why identical steps in each language are printed on facing pages of all checklists. This dual-language operation is also reflected in Mission Control procedures, crew timelines, and time-critical and safety-critical contingency operations.

In addition to ensuring clear dual-language procedures, the Russian Liaison astronaut also facilitates coordination between the two agencies by developing strong working relationships between control teams. He/she also attends practices and simulations to ensure unambiguous communications and methods are used during pre-mission training in addition to developing procedures for exchanging information between various control centers. Often, this requires the Russian Liaison astronaut to attend technical discussions and meetings, sometimes requiring the use of an interpreter when verbal exchanges become linguistically challenging!

Crew Support Astronaut

A more popular assignment is serving as the Crew Support Astronaut, who acts as the primary contact for all crew needs, coordination, planning, and interactions. He/she is also the primary representative of the crews while they are on orbit, which means dealing with all sorts of issues ranging from ensuring personal items are loaded onto the resupply capsule to a death in the family. For example, in December, 2007, astronaut Dan Tani's mother died while he was onboard the ISS. Although the task of informing Tani was performed by Dr Sean Roden, Tani's flight surgeon, the Crew Support Astronaut played a role liaising with family, friends, and counselors.

The Crew Support Astronaut also works with NASA's Flight Crew Equipment Representative (FCER), ensuring flight crews have all their personal items when they fly to the ISS. This might sound like a simple job, but everything an astronaut takes into space must conform to NASA guidelines, which inevitably means wading through more paperwork! If an item doesn't comply with NASA regulations (which sometimes happens with souvenir items), the Crew Support Astronaut might work with the FCER to find a way to approve the item.

Public speaking

In addition to their technical assignments, astronauts must also perform collateral duties, one of which is giving public speeches and press interviews. Since the Astronaut Office receives hundreds of requests every month for speakers, astronauts often have a selection of venues to choose from. Canadian Space Agency (CSA) astronauts, for example, usually spend at least 2 days every month visiting schools and colleges around Canada. For the space agencies, these so-called "blue suit" (the astronauts usually wear their blue flight suits to public speaking events, hence the term) public relations events generate a lot of visibility with the taxpayer as well as generating interest at the grassroots level. For some of the astronauts, the PR circuit is sometimes a series of events guaranteed to increase their heart rates, as they try to steer a steady line of political correctness and provide insightful answers to questions such as "Do you believe in aliens?", "How many gay astronauts are there?", and "Have astronauts had sex in space?". More often, the questions are a little more serious, with journalists asking about science and technology and the space program in general. These types of questions not only provide the astronauts a welcome opportunity to put their employer in a good light, but also an opportunity to talk about their favorite subject: manned spaceflight.

A typical technical assignment may last a year and, depending on the number of flights available, an astronaut may perform three or more assignments before receiving their flight assignment. Not surprisingly, while astronauts welcome the opportunity to work behind the scenes, not a day goes by without them thinking about their first flight into space. Of course, before they actually get to fly, astronauts must complete even more training! But, before we look at their final phase of training, it is worthwhile knowing what types of missions today's astronauts will be assigned to.

6

Types of missions

In between the myriad technical assignments and public relations work, astronauts inevitably spend time thinking about the day they will receive the coveted mission assignment. Until recently, astronauts were selected for either a Space Shuttle mission or an International Space Station (ISS) increment. However, with the retirement of the Shuttle in 2010, the only mission available for new astronauts will be Expedition Class increments onboard the ISS. This is because NASA's new family of launch vehicles and crew exploration vehicle (CEV) will not be operational until 2015 – at the earliest! In the interim, NASA, the Canadian Space Agency (CSA), and the European Space Agency (ESA) astronauts will hitch rides onboard the Russian Soyuz vehicle at a cost of $51 million a seat!

The retirement of the Shuttle will release much needed funds NASA will use to develop a new family of launch vehicles that are needed to return humans to the Moon. NASA's return to the Moon is planned for 2020, after which a series of missions will be launched to the lunar surface, with plans to establish an outpost by 2025. Once the outpost is established, NASA will turn its sights towards a manned mission to Mars. NASA and CSA astronauts selected in the 2009 class may fly one or two missions to the ISS followed by a short or long-duration mission to the Moon. For those lucky enough (and young enough!) not to have exceeded their career radiation limits, it is possible they may be selected for a Mars mission. This chapter provides an insight into each of these types of mission.

INTERNATIONAL SPACE STATION MISSIONS

The following is a synopsized account of the last 4 months of ISS Expedition 17 (Figure 6.1), whose members included Russian Federal Space Agency cosmonauts Sergei Volkov (commander), Oleg Kononenko (flight engineer), and NASA astronaut Greg Chamitoff (flight engineer).

Figure 6.1. International Space Station Expedition 17 Crew. Russian Federal Space Agency cosmonauts, Sergei Volkov (center), Expedition 17 commander; Oleg Kononenko, flight engineer; and NASA astronaut, Greg Chamitoff (left), flight engineer, take a break from training at NASA's Johnson Space Center to pose for a crew portrait. Image courtesy: NASA.

Expedition 17: July 23rd to end of November, 2008

After conducting EVAs on July 10th and 15th, the Expedition 17 crew returned to routine ISS operations (Table 6.1) involving maintenance, exercises, and experiments. On July 23rd, the crew initiated a burn to raise the orbit of the complex (Figure 6.2) by 7.21 km to approximately 351 km. Following the burn, the crew conducted routine maintenance work on the Russian segment, which included replacing a condensate separation and pumping unit in the Zvezda condensate water processing system. On July 27th, Volkov and Kononenko recharged the Zvezda's SKV-2 air conditioner with coolant from a bottle delivered by a Progress craft (the air conditioning loop had been switched off since a coolant spill in April). In early August, Kononenko removed electronic components from the Zarya module's solar array positioning mechanism, after which the crew went to work replacing one of the hoses on the Zvezda toilet, following a malfunction of the unit. In between the plumbing work, the crew performed Soyuz descent training and emergency evacuation drills. The crew also managed to catch up on their science schedule. For example, Chamitoff was charged with conducting an acoustics experiment requiring him to place acoustic dosimeters in strategic locations throughout the ISS, such as sleep locations and work areas. Chamitoff was also responsible for conducting the Sleep-Long and Light Exposure investigation, requiring him to wear an Actiwatch and write entries in a sleep log describing the quality and duration of his sleep.

Table 6.1. Typical ISS routine (September 25th, 2008).

GMT	*Crew*	*Activity*
06.00–06.10	FE-1, FE-2	Morning inspection
06.00–06.10	CDR	SONOCARD. Experiment completion
06.10–06.40	FE-2	Post-sleep
06.10–06.20	CDR	Morning inspection
06.10–06.40	FE-1	Post-sleep
06.20–06.40	CDR	
06.40–07.30	All	Morning meal
07.30–07.40	All	Work prep
07.40–07.55	All	Daily planning conference (S-band)
07.55–08.00	FE-2	LAB window closing
07.55–08.15	CDR, FE-1	Work Prep
08.00–08.05	FE-2	JEM window closing
08.05–09.35	FE-2	FE RED
08.15–08.45	CDR	БСК-2В device teardown. *Conf with specialists if needed (S-band)*
08.45–09.15	CDR	Routing cables – 17КС.10Ю 8210А-3730, -3740, -3860 БСК-1В to connect АСН-М (SatNav system). *Conf with specialists if needed (S-band)*
08.50–10.20	FE-1	FE TVIS-4
09.15–09.30	CDR	SONOCARD. Copying data to Laptop RSE-MED
09.30–09.40	CDR	Filtration unit C02 ГА ИК0501 R&R. Install unit #30 (CM1_4_449_1, barcode 008545R , bag 363-25); replaced unit is planned for disposal
09.35–10.35	FE-2	FE CEVIS
09.50–10.20	CDR	Turn off PRINTER1. Install БСК-2В device to its nominal location (after АСН-М (SatNav system) test). *Conf with specialists if needed (S-band)*
10.20–11.50	CDR	FE TVIS-4
10.35–10.55	FE-2	In-Flight Maintenance CEVIS
10.55–11.30	FE-2	OUM-PFE – h/w installation
11.20–12.00	FE-1	Laptop antivirus database update (for RSS1, RSK1, RSK2, RSE1). *Conf with specialists (S-band)*/rg
11.30–11.35	FE-2	Ham radio h/w prep
11.35–11.45	FE-2	Ham radio comm. session

Table 6.1. *continued*

GMT	*Crew*	*Activity*
11.45–12.15	FE-2	OUM-PFE – installation
11.50–12.50	CDR	Midday meal
12.00–12.15	FE-1	ГАНК data readout
12.15–13.15	FE-1,FE-2	Обед
12.50–13.00	CDR	Activating PRINTER1 in SM. *Conf with specia lists if needed (S-band)*
13.15–14.15	CDR	Cardio-ODNT. Experiment preparation
13.15–13.25	FE-2	Systems DC converter and microgravity measurement equipment activation
13.15–14.15	FE-1	Cardio-ODNT. Experiment preparation (for assisting crew member)
13.25–14.25	FE-2	Locating and installing K-bar h/w in *Columbus*
14.15–14.35	CDR,FE-1	Cardio-ODNT. Measurements during comm. pass To = 14.16. *Conf with specialists (VHF)*
14.25–14.45	FE-2	Photographing Lab-On-a-Chip glass slide
14.45–15.15	FE-2	Updating PCS computers with software version R11
15.05–15.45	CDR	СОЖ, Maintenance
15.05–15.45	FE-1	"Electron" activation and Secondary Purification Unit temp monitoring. *Conf with specialists if needed (S-band)*
15.15–16.15	FE-2	PCS hard drives replacement
15.45–15.55	FE-1	New Ethernet clients comm. check
15.45–17.45	CDR	СУ-95 matching unit installation in DC1. Install unit 002 (ФГБ1ПГО_4_417_1, barcode 00045690R, soft container 364-8) *Conf with specialists (S-band)*
15.55–16.55	FE-1	FE TVIS-4
16.15–17.35	FE-2	Stowing SODF
16.55–18.40	FE-1	Stowing SODF
17.35–17.45	FE-2	CWC inventory
17.45–18.00	FE-2	Starting filling CWC from LAB condensate container
17.45–18.05	CDR	IMS update
18.00–18.30	FE-2	Work prep
18.15–18.45	CDR	
18.30–18.45	FE-2	Finishing filling CWC from LAB condensate container
18.40–18.45	FE-1	TVIS RED HRM data transfer to MEC
18.45–19.00		Daily planning conference *(S-band)*
19.00–19.30		Work prep
19.30–20.00		Evening meal
20.00–20.30		Daily food prep
20.30–21.30		Pre-sleep
21.30–06.00		Sleep

In August, Chamitoff completed a series of tests as part of NASAs Nutrition experiment, designed to assess bone metabolism, oxidative damage, and hormonal changes during long-duration spaceflight. The tests included collecting urine samples

Figure 6.2. International Space Station. Regular re-boosts are needed to overcome the effects of residual atmospheric drag, which makes the station lose about 100 m in altitude per day. These re-boosts are usually performed by the European Space Agency's Automated Transfer Vehicle located to the left in this image. Image courtesy: NASA.

and banking them, together with a series of blood samples that were made available to investigators post flight. In addition to the tests, Chamitoff completed a regular series of questionnaires chronicling the effects of isolation and confinement onboard the ISS. In between conducting experiments and maintenance activities, the crew continued to use the ham radio to make contact with schools for educational events as part of the Amateur Radio on the International Space Station (ARISS) experiment. Other activities included taking images as part of the Crew Earth Observation program. For example, the crew captured images of Tropical Storm *Edouard* as it headed through the Gulf of Mexico (Figure 6.3).

While Chamitoff was hard at work conducting NASA's science experiments, Volkov and Kononenko continued the Russian experiment program. In addition to the medical experiments, which included bi-weekly calf volume and body mass measurements, and blood and urine analysis, the two cosmonauts conducted the Profilaktika (Russian for "prevention") experiment. The investigation, which measures the effects of microgravity on the human body, required the cosmonauts to ride a stationary bike while wearing a breathing mask and heart rate monitor. A

Figure 6.3. Tropical Storm *Edouard* captured from the International Space Station as it headed across the Gulf of Mexico during Expedition 17. The storm caused Johnson Space Center to be closed between 4th and 6th of August, 2008. Image courtesy: NASA.

few days later, the pair conducted the Cardio-ODNT, which investigates responses of the blood circulation when negative pressure is applied to the lower part of the body. In the second week of August, Volkov and Kononenko once more turned their hand to maintenance activities when they repaired a torn tread belt on the Treadmill with Vibration Isolation System (TVIS). They also replaced roller bearing assemblies on the TVIS before checking it out and allowing it to be used for exercise once again.

While the cosmonauts worked on the treadmill, Chamitoff spent some of his off-duty time playing "space chess" with ground controllers (Figures 6.4 and 6.5). The "space chess" was a team-building exercise involving several NASA centers. Each day, Chamitoff would play against ground controllers at a specific NASA center before moving on to the next center the following day. The off-world chess tournament started on August 13th, with Chamitoff winning the first game.

A week after the chess tournament came to a close, the crew was busy scanning their laptop computers after the discovery of a virus. It was thought the virus, which may have been transferred to the laptops from a flash drive (this is one of the reasons the ISS does not have direct access to the internet), was intended to steal passwords and relay them to a remote server.

After resolving the virus problem, the crew turned its attention to two of the science racks inside the American Destiny Laboratory. By the end of September, nine of Destiny's racks were planned for relocation into the Japanese *Kibo* module.

Figure 6.4. Greg Chamitoff considers his strategy carefully as he prepares for his next move against flight control teams. Image courtesy: NASA.

Figure 6.5. Flight Director, Chris Edelen, considers a move he hopes will stop Earth's losing streak in the space chess competition. Image courtesy: NASA.

Figure 6.6. Chamitoff works in the Kibo laboratory to move an experiment rack during a relocation task. Image courtesy: NASA.

Removing the racks was a laborious task (Figure 6.6), requiring the crew to disconnect associated laptops, cables, and support equipment before floating the racks into *Kibo*, where they had to reconnect everything again. Later in the week, the crew ignited the engines of the Automated Transfer Vehicle (ATV) for 5 min to lower the orbit of the ISS. The firing was a Debris Avoidance Manoeuvre (DAM) to take the ISS out of the path of a possible collision with a small piece of debris from a Russian satellite.

In the first week of September, the crew made preparations to bid farewell to the ATV. The ATV had brought 269 liters of water, 21 kg of air and 811 kg of refuelling propellant to the ISS in June and now it was time to use the vehicle as a waste disposal unit. It took the crew 31 hr to load dry cargo and liquid waste (a cumulative total of more than 1,000 kg) into the ATV. Once the ATV was loaded with material, the crew conducted a safety check, closed the hatch between the ISS and the ATV, and undocked the vehicle from the rear Zvezda docking port. After performing a burn of its attitude thrusters, the ATV gradually increased its distance (Figure 6.7) from the ISS until it reached an altitude 5 km below the station, at which point the crew disabled the station's automatic emergency systems (which can initiate a collision avoidance maneuver).

Next on the crew's agenda was preparing for the arrival of a Progress cargo vehicle. On September 17th, the crew, aided by Mission Control in Russia and NASA's backup (JSC was closed during the docking due to a hurricane) control

Figure 6.7. The ESA's Automated Transfer Vehicle (ATV) begins its relative separation from the International Space Station. Image courtesy: ESA.

team in Austin, Texas, commanded the Progress vehicle to a docking with the aft port of the Zvezda module. Shortly after, the crew was inside the cargo craft unloading science equipment and personal items. A week after the docking of the Progress vehicle, the crew began relocating racks inside the American module. The relocation was needed to increase the science capabilities of the European *Columbus* and Japanese *Kibo* laboratories and to create room for new life support racks. The first to be moved was the Crew Health Care Systems (CHeCS) rack, followed a week later by the Human Research Facility (HRF) rack. Once again, the crew had to disconnect a snake's nest of umbilicals before reconfiguring the connections after the racks had been relocated.

In between the moving of science racks, the crew continued with its experiments. Chamitoff began the Sodium Loading in Microgravity (SOLO) experiment, a new life-sciences study requiring him to follow a rather unpalatable high-salt diet to investigate the mechanisms of fluid and salt retention in the body during long-duration flights. Meanwhile, the cosmonauts, in preparation for their return to Earth on October 24th, conducted regular sessions in the Chibis low-pressure leggings. Unfortunately, their preparation for return to Earth was interrupted when the ISS toilet broke down after the failure of a gas separator, requiring Kononenko to practice his plumbing skills. Once the toilet was fixed, the crew turned its attention to the arrival of Soyuz TMA-13, carrying Volkov's and Kononenko's replacements. The Soyuz docked with the ISS on October 14th, carrying Yuri Lonchakov, Michael

Fincke, and the world's sixth spaceflight participant, Richard Garriott. Shortly after the docking, the two crews commenced handover work with Volkov and Kononenko briefing Fincke and Lonchakov on the detailed operation of the myriad systems in the ISS. The following week, the crew conducted a change of command ceremony and Volkov and Kononenko made final preparations for leaving the ISS, packing away samples in the descent module and running through descent procedures. On October 24th, the Soyuz undocked from the ISS and Volkov fired the Soyuz engines to increase the distance from the station. Shortly thereafter, the Soyuz began a nominal entry into the Earth's atmosphere, followed by a landing in Kazakhstan, 90 km away from the city of Arkaykh. Volkov and Kononenko had spent 198 days, 16 hr and 19 min in flight.

Following the departure of Volkov and Kononenko, Fincke, Chamitoff, and Lonchakov had 3 weeks to prepare the ISS for the arrival of STS-126. During their preparations, Fincke and Chamitoff participated in their first Periodic Fitness Evaluation test, taking blood pressure and electrocardiograph (ECG) measurements while exercising on the Cycle Ergometer with Vibration Isolation System (CEVIS). On October 29th, the engines of the Progress vehicle were fired to boost the station's altitude by 1 km to 352 km, thereby placing the ISS in the correct orbit for the arrival of the Space Shuttle and the next Progress craft. In addition to the science work and boosting the station's altitude, the crew also performed the usual shopping list of maintenance activities, ranging from servicing the recalcitrant Zvezda toilet to exchanging diffuser plates on the Temperature and Humidity Control (THC) system in the Harmony node. In mid-November, the crew once again had to take out the garbage. After packing the Progress vehicle with unneeded items and rubbish, they commanded the undocking of the craft before firing its engines, causing it to descend 40 km per orbit. The next day, the Space Shuttle, with a crew of seven, launched from KSC, carrying 6,500 kg of cargo, enabling the ISS crew to double from three to six members in June, 2009.

While the Space Shuttle was berthed with the ISS, several EVAs (Figure 6.8) were conducted to install new hardware, such as the Lightweight Multi-Purpose Experiment Support Structure Carrier (LMESSC), and effect repairs, such as replacing the trundle bearing assemblies (TBA) for the Solar Alpha Rotary Joint (SARJ). While some astronauts floated about outside, astronauts inside the station were busy working with the much publicized Water Recovery System (WRS) after the Urine Processing Assembly (UPA) caused it to shut down, much to the embarrassment of NASA. After a marathon series of EVAs and ISS maintenance, the Shuttle and ISS crews said their farewells on Thanksgiving Day and *Endeavour* undocked, leaving former STS-126 crewmember Sandra Magnus on the ISS as Chamitoff's replacement.

While ISS increments will generate important information regarding the performance of humans during long-duration missions, the station serves primarily as a stepping-stone for missions to the Moon and beyond. While astronauts will still undoubtedly savor their time onboard the ISS, many will be looking forward to the time when humans finally return to the Moon. What they will do once they get there is described in the following section.

Figure 6.8. Astronauts conduct one of several extravehicular activities during Expedition 17. Image courtesy: NASA.

LUNAR MISSIONS

Surface activities

The core of the Vision for Space Exploration (VSE) is exploration. To achieve exploration goals, astronauts will search for resources, learn how to work safely in a harsh environment, and explore the lunar surface. These activities (Table 6.2) will require sustained periods of EVA outside the protective environment of the outpost.

Table 6.2. EVA tasks during lunar missions.[1]

EVA	*Description*	*EVA*	*Description*
Site preparation	• Survey and stake-out • Rock removal • Smoothing • Clearing dust control areas • Establish navigation aids	Habitat installation	• Hole excavation • Transport hab modules to base site • Unload, locate, level • Deploy and inflate back filling
Shielding installation	• Support for regolith • Regolith bagging for radiation protection • Bag stacking and lifting clearing access paths	Science	• Sample collection • Installation of experiments • Location of mapping for geological survey • Establish observatories
Power systems and thermal control system	• Site preparation • Unload equipment from lander and transport • Deploy and assemble • Radiator deployment and activation • Connect to distribution • Shielding excavation	Logistics	• Unload from hangar • Unpack • Transport • Transfer • Storage • Waste disposal • Storage of spent recyclables
Resource operations	• Resource process site set-up (pressure vessel, plumbing, gas holding tanks, pumps, heat exchangers)	Lander operations	• Servicing/minor repairs • Refueling • Pre-launch and checkout • Relocation of fuelling depot
Mining operations	• Equipment transportation to site • System set-up (bucket wheel excavator, conveyors, regolith bagging equipment, sorters, separators) • System relocation	Upkeep	• Inspection • Field checks and measurements • Replacement of systems/subsystems • Repair of equipment

Bio-Suit

EVAs will require astronauts to wear an extra-vehicular mobility suit (EMU), comprising a spacesuit assembly and a portable life support system (PLSS). The hostile environment of the lunar surface, combined with the challenges posed by the conditions of reduced lunar gravity and the duration of surface excursions, will require a tremendously versatile and rugged suit. Such a suit is the Bio-Suit (Panel 6.1), a concept that may eventually revolutionize human space exploration. The Bio-Suit is based on the concept of biomechanically and cybernetically augmenting human performance capacity (Figure 6.9). It was conceived by Professor Dava Newman of the Massachusetts Institute of Technology (MIT), who envisioned the suit to function as a second skin by providing mechanical counter-pressure (MCP).

Panel 6.1. Bio-Suit

Integrated into the Bio-Suit is a bioinstrumentation system modeled on the Operational Bioinstrumentation System (OBS), developed in the 1970s at the University of Denver for use during Space Shuttle missions. The original OBS consisted of a Signal Conditioner, an EVA Cable, a Sternal Harness, and three electrodes – a suite of components that will be significantly upgraded in the Bio-Suit, which will feature biosensors integrated into the textiles. The biosensors will measure biomedical signals such as heart functions, oxygen consumption, and body temperature, while a suite of biochemical sensors will provide information concerning body fluids and dosimeters will measure the local radiation environment.

An EVA system fitted with wearable sensors becomes increasingly important as surface excursions become longer, EVA locations become more isolated, and activities become increasingly complex. The combination of these factors has the potential to increase the distraction and fatigue of the crewmember. However, should an accident occur in an isolated location or a crewmember become excessively tired, having access to the information provided by a wearable sensor system may improve chances of survival.

Whereas the lightest NASA EMU weighs at least 40 kg, Newman's skin-tight Bio-Suit will weigh two or three orders of magnitude less. Constructed of spandex and nylon, the multi-layered suit hugs the body's contours like a second layer of skin and its MCP technology ensures constant pressure is applied to the surface of the body. This pressure is needed not only to counteract the vacuum of the lunar surface and maintain the body's homeostasis, but also to avoid blood pooling.

Maintaining an even pressure over the surface of the human body has, until now, been achieved by utilizing the bulky gas-pressurization systems embodied by NASA's bulky EMU suits. Thanks to new MCP technology that works along lines

Figure 6.9. Future lunar attire: the Bio-Suit. Image courtesy: Professor Dava Newman, MIT: inventor, Science and Engineering; Guillermo Trotti, A.I.A., Trotti & Associates, Inc. (Cambridge, MA): design; Dainese (Vicenza, Italy): fabrication; Douglas Sonders: photography.

of non-extension (those lines along the body undergoing little stretching as the body moves), the gas-pressurization system of the H-Suit is no longer necessary. Another advantage of the Bio-Suit is suit compromise; if NASA's traditional EMU suit suffers a puncture, the crewmember must return to the outpost and undergo decompression, whereas a small puncture of the skin of the Bio-Suit will require only a bandage!

Science on the Moon

Clad in their Bio-Suits, astronauts embarked upon lunar outpost missions will spend a significant amount of their time conducting scientific activities such as the ones described here.

Astronomy and astrophysics

194,068 square km on the far side of the Moon have been set aside for the purpose of establishing a future Moon base for observatories. One of these, the Icarus Lunar Observatory Base (ILOB), will be located at the point farthest away from Earth and, thanks to the absence of atmospheric interference and sunlight, provide astronomers with the perfect location to establish optical and radio telescopes. In addition to requiring astronauts to assemble and emplace the ILOB, astronauts will also be required to remove the lunar dust that, charged by the solar wind, will attach electrostatically to the components of the observatory.

Of particular importance to those living on the Moon will be studies aimed at defining, characterizing, and predicting the lunar radiation environment. Since the Moon is outside the Earth's magnetosphere and lacks an atmosphere, it is subject to constant bombardment by energetic solar particles and cosmic rays. By installing large high-energy cosmic ray detector arrays on the lunar surface, astrophysicists will be able to measure radiation striking the surface and perhaps provide a method to forecast serious radiation events. Once again, the emplacement of such equipment will be the task of astronauts living on the surface.

Earth observation

The surface of the Moon is the ideal location for a remote sensing platform, since it affords global observation of the Earth. Synthetic Aperture Radar (SAR) will utilize this feature to create highly accurate terrestrial topography, altimetry, and vegetation charts. The lunar surface also provides the perfect vantage point for simultaneous observations of the Earth–Sun system, helping scientists to understand the reaction of the Earth's atmosphere to solar activity, in turn helping to estimate the effect of long-term solar changes on climate. The advantage of viewing the whole Earth disc also enables the collation of information concerning land surface mineralogy, land use, land change, and biomass utilization. Once again, astronauts will be instrumental in deploying and maintaining the equipment required to help Earth-bound scientists achieve these objectives.

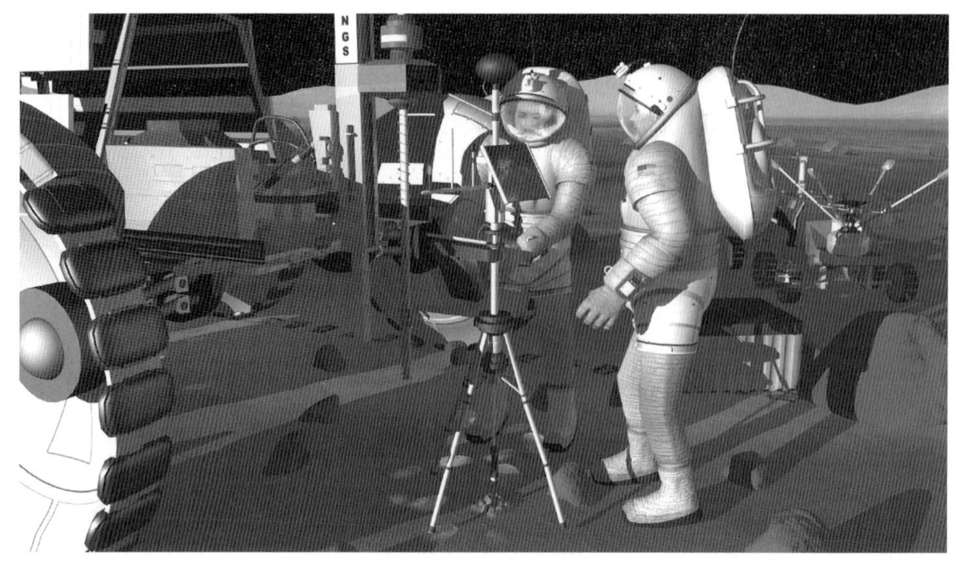

Figure 6.10. Artist's rendering of astronaut conducting seismic exploration on the Moon. Image courtesy: NASA.

Geology

One of the first geological investigations performed by astronauts may be collating data on Moonquakes, providing lunar inhabitants with a means to determine which areas are safe and which are liable to seismic activity (Figure 6.10). In addition to revealing site hazards, seismic exploration will also help geologists define mantle characteristics, identify buried lava tubes and provide an insight into the processes involved in planetary evolution and crustal genesis.

Other geological variables of interest will include the characterization of endogenous (from the lunar interior) and exogenous (delivered externally) processes, which resulted in the deposition of volatiles on the lunar surface. These studies will lead to predictive models of the distribution of volatiles that may assist in the procurement of materials for in-situ resource utilization (ISRU). Similar studies will be directed at characterizing the general geology of the Moon by investigating materials present in regolith to locate and identify elements useful in the development of a lunar outpost. Accomplishing these studies will require astronauts to use the lunar rover (Figure 6.11) to visit sites of interest.

In addition to conducting geological exploration, astronauts will also investigate the lunar regolith. These investigations will be necessary not only to define potential ISRU sites, but also to help understand the nature and history of solar emissions, galactic cosmic rays (GCRs), and dust from interstellar medium, all of which are preserved in lunar regolith. The study of regolith will also provide information permitting geologists to better characterize the process of space weathering of

Figure 6.11. One iteration of NASA's lunar rover. Image courtesy: NASA.

planetary bodies without air or water. These studies will require astronauts to map regolith maturity to identify regions of ancient and new regolith. These data will then be used to understand the weathering process and also provide a reference point for understanding the potential hazards of resource mining on the Moon.

Materials science

One of the primary efforts of material scientists will be to design mitigation strategies aimed at ensuring the robust performance of hardware, particularly during extended stays of 6 months or more. To achieve this, astronauts will deploy investigations to study processes such as fractional gravity, ultra-high vacuum, radiation bombardment, thermal cycling, and perhaps the lunar inhabitants' biggest headache – lunar dust. Characterizing the cumulative effects of these aspects of the lunar environment will enable engineers to make informed decisions regarding the materials best suited to withstand extended use on the Moon.

Physiological adaptation

Every astronaut traveling to the Moon in the next several decades will serve as a lab rat for terrestrial physiologists tasked with understanding the effects of extended lunar stays and ensuring the health and safety of crewmembers. While the effects of fractional gravity upon the body are somewhat understood, others, such as the effects of deep space radiation bombardment, will require further study. In addition to characterizing the physiological response to extended lunar stays, it will be necessary to study human factors, such as strategies to help astronauts deal with the effects of prolonged isolation, and communication lag on performance and mission coordination. These data will be accumulated gradually by astronauts during their surface stays and, post mission, by scientists examining returning crewmembers.

In-situ resource utilization

ISRU is a process involving any hardware or operation exploiting and utilizing either natural or discarded in-situ resources to create products and services for robotic and human exploration. For example, on the Moon, *natural* in-situ resources include regolith, minerals, volatiles, metals, water, sunlight, and thermal gradients, whereas *discarded* in-situ materials include the descent stage of the LSAM, fuel tanks, and crew trash.

One of the first ISRU missions astronauts will be tasked with will be ensuring their own survival. This will require protecting themselves from the harsh radiation environment by using regolith to bury the habitat. Once the habitat becomes established and ISRU processes become more developed, the crew will begin to process lunar materials to produce water and oxygen, used as a contingency supply for the Environmental Control Life Support System (ECLSS) and as a consumable source for EVA missions.

Extracting water ice

Analysis of lunar samples returned by the Apollo missions confirm lunar soil is rich in oxygen, silicon, iron, calcium, magnesium, and titanium, each of which may be extracted from the regolith by various processing techniques. In addition to conducting various ISRU studies, the astronauts may be tasked with extracting water from the ice-rich lunar regolith found in the cold trap areas near the lunar South Pole. The latest analysis of data from Lunar Prospector's neutron spectrometer indicates there may be as much as 300 million tonnes of water ice contained in the permanently shadowed cold traps near the poles. To extract water ice, astronauts may use a regolith reduction reactor, which will work by simply heating the regolith, using solar energy.

Developing construction materials

As the outpost grows, there will be an increased need for construction materials. Rather than pay the exorbitant cost of ferrying these materials from Earth, it will make more sense to use what is available locally. One method the astronauts may use to develop construction materials from the lunar regolith is *sintering*. Sintering is a process in which soil is heated and maintained at temperatures below its melting point, thereby permitting viscous flow on the grain surfaces to form necks between particles, resulting in the particles bonding together. Studies investigating sintering of lunar soil[2] returned by the Apollo 14 mission demonstrated that the process results in a porous and brittle material, although the cohesiveness of the solid can be increased by applying higher temperatures and greater pressures. No doubt, astronauts will be heavily involved in testing these procedures; in doing so, they will be able to add "off-world construction worker" to their long list of qualifications!

Mining Helium-3

In addition to their science, water extraction, and sintering activities, astronauts will also be involved in mining. In 1985, student engineers at the University of Wisconsin discovered that a sample of lunar soil, taken by Apollo 17 astronaut-scientist, Harrison Schmitt, from the rim of the Moon's Camelot crater, contained significant quantities of a special type of helium known as Helium-3 (He-3). The discovery intrigued the scientific community due to He-3's unique atomic structure, making it a perfect fuel for nuclear fusion, a process capable of generating considerable amounts of electrical power. In addition to being extremely potent, He-3 is non-polluting and has virtually no radioactive by-product, making it an ideal candidate for a twenty-first-century fuel source. Unfortunately, hardly any He-3 exists on Earth but scientists estimate there may be a million tonnes of He-3 on the Moon, enough to supply the world's energy requirements for several thousand years.

He-3 is deposited in the powdery lunar soil when solar wind, composed of a stream of charged particles emitted by the Sun, strikes the Moon. This process has continued for billions of years, resulting in a potential cash crop of He-3 ready to be strip-mined from the lunar surface. Extracting He-3 will require teams of astronauts trained in the art of mining and the handling of robotic equipment. Once refined, the He-3, which would now be in the form of either pressurized gas or cryogenic liquid, would be ready to be shipped to Earth by unmanned spacecraft fuelled by lunar-derived propellant systems.

Oxygen production

To ensure the settling and evolution of a permanent and autonomous lunar outpost, astronauts will need to produce lunar liquid oxygen (LLOX) utilizing indigenous resources. Not only will such in-situ utilization result in cost savings on propellant for transportation systems, but it will also support the astronauts living on the lunar surface. Fortunately, oxygen is the most abundant element on the Moon but the problem is it must be extracted from lunar rock and regolith, where it exists in chemical combination with elements such as iron and titanium.

Thermal recovery of oxygen

One simple process astronauts may use to produce oxygen is to roast lunar soil to temperatures of 900°C.[3] At this temperature, all the hydrogen contained in the lunar soil is released and can react with oxide minerals to produce water in a process similar to that used to reduce ilmenite by hydrogen. Electrolysis may then be used to dissociate water into hydrogen and oxygen. Although the process is simple and requires no difficult development, a major disadvantage is the huge quantity of lunar soil that would require processing. However, this shortcoming may be slightly offset by the possibility of co-producing oxygen, helium, and hydrogen in the recovery process.

If all goes to plan, NASA (possibly with some help from the ESA) will have established a lunar outpost by 2025. Just as the ISS served as a test-bed for the technologies used for a lunar outpost, so too will the Moon base serve as a launch pad for missions to Mars.

MARS MISSION

A journey to Mars will rival the journeys of Shackleton, Amundsen, and Nansen, especially in terms of mission duration, and isolation from sources of supply and assistance. Once in transit from Earth to Mars, the crew will have no resupply and only limited resources and capabilities available to them to maintain and repair the vehicle. While the ISS is outfitted with a lifeboat to assure crew safety in case a system fails or astronaut illness warrants a return to Earth, astronauts en route to Mars will have no such luxury. Once the vehicle performs its trans-Mars insertion (TMI) burn, the crew will be committed to a trajectory to the Red Planet, and support from Earth will be limited to communications and ground-based experience.

Crew activities en route

Once en route, one of the first daily rituals of the pilot will be checking the trajectory of the vehicle as it heads for Mars. If the flight path is found to be misaligned, it will be necessary to make a trajectory correction maneuver (TCM) based on corrective burn parameters calculated by Mission Control and also information provided by the inertial navigation system (INS) onboard the vehicle. Having checked they are still on course, the crew will then settle down to daily life onboard the habitation module.

Contingencies

Given the seemingly endless list of possible contingencies, it is inevitable the crew will become familiar with dealing with emergencies. Some off-nominal events will be benign, others false alarms, and most will be non-life-threatening, but some may have the potential to threaten the crew and the mission.

Perhaps the most worrying of all the threats is that posed by a solar particle event (SPE), a burst of potentially lethal radiation that will trigger the automatic flare alarm. Although the spacecraft will be designed to shield the crew from solar and galactic radiation, the intense doses resulting from a SPE may be a life-threatening event. During the SPE, astronauts will retreat to the storm shelter onboard the spacecraft and spend much of their time closely monitoring their personal dosimeters, to gauge how much radiation their bodies are absorbing.

Expedition clothes

The crew will probably spend most of their time wearing a fire-retardant, two-garment ensemble, consisting of pants and shirts. How long they will wear the same set of clothes is open to question, as there is no documented interplanetary mission protocol for clothing changes. Since there will be no washing facilities onboard, crewmembers will bring several sets of clothing with them but, due to space restrictions, there will obviously be a limit. For example, weekly changes of their mission uniform would total more than 300 items and would impose a significant weight penalty, so it is likely crewmembers will only change twice per month.

Hygiene

Hygiene standards are likely to contrast markedly with those deemed acceptable on current ISS missions. Although astronauts will still shave, trim their nails, brush their teeth, and wash their faces and hands regularly, the daily shower will either be a luxury activity or unavailable, due to space restrictions. Given the cramped quarters of the crew vehicle, it is likely a collapsible shower will be provided, requiring crewmembers to unstow and deploy the shower once or twice a week.

Running to Mars

Although the recreational value of exercise will no doubt be acknowledged by Mars-bound astronauts, the requirement to spend 3 or 4 hr running or cycling every day will also have other benefits for those confined to a microgravity environment. Evidence from dozens of long-duration space missions has confirmed astronauts subjected to long periods of microgravity lose not only bone mass, but also muscle strength. The loss of bone mass is the greatest concern for mission planners, since the bones of astronauts could become so weak they could fracture under Martian gravity. Astronauts assigned to long-duration increments onboard the ISS understand the insidious effects of bone demineralization and most adhere to a strict regime of exercise involving at least 2 hr of running on a treadmill or cycling on an ergometer. Even when following such a rigorous training regime, astronauts returning to Earth still lose as much as 10% of their lower body bone mass.

Preparing meals

Exploration missions and prolonged periods of zero-gravity present challenges for food preparation. The crew will not have a refrigerator or freezer onboard, so foods will be thermo-stabilized, irradiated, freeze-dried, or canned. Due to the absence of friends, family, and normal recreational pursuits, combined with the effects of prolonged isolation and confinement, food will assume added importance, as crewmembers will focus on meals as a substitute for the customary sources of personal gratification on Earth. It is likely the preparation of meals will be viewed as a pleasant activity and distraction from the otherwise mundane routine of exercise and inflight monitoring of data. Also, it is probable some astronauts will take advantage of the low-tempo operations and spend time preparing labor-intensive meals for their fellow crewmembers.

Working en route

Given the duration of a Mars expedition, mission planners will have to work hard to ensure boredom does not become a significant mission stressor. To do this, workloads will be devised to ensure mission-related tasks are divided equally among crewmembers. Tasks will probably include regular contingency simulations, activities preparing for landing on Mars, reviews of descent procedures, and, of course, exercise. Yet, even when following a carefully scripted day-to-day list of assignments, astronauts will inevitably occasionally find themselves with nothing to do.

Personal communication

During the mission, news will not travel as quickly as it did on previous LEO missions. Unlike all space missions to date, astronauts en route to Mars will not have the luxury of two-way communication due to the lag time, which will increase as the mission progresses. In fact, by the time the astronauts arrive at Mars, the lag time may be as much as 20 min. Due to the response lag, crewmembers will instead log personal communication messages from their private quarters. The messages will then be sent via a scrambler to ensure transmissions are kept private.

Getting along

The job of the interplanetary astronaut will be a stressful one. Fortunately, astronauts are trained to deal with stress better than almost anyone. Not only must they deal with the dangers of explosive decompression, slow death by radiation sickness, and equipment malfunctions, they must also deal with each other, living in close-quarters 24 hr a day for month after month. For a crew bound for Mars, the stress will be compounded by mission complexity, limited abort options, and the sheer length of time away from home and family. Unsurprisingly, the effect of being cooped up with other members in what constitutes a tiny society for several months will exert additional stress upon crewmembers. To offset interplanetary strains and anxieties, crewmembers will need to find ways to relax.

Leisure time

An insight into the type of leisure activities astronauts might pursue has been gleaned from research conducted at Antarctic research stations, onboard nuclear submarines, in accounts of polar explorers, and, of course, from previous long-duration space missions.

Based on these reports, one of the most popular off-duty activities is simply talking, with a tendency for reading to occupy more time as a mission progresses.[4, 5] Also high up on the list is watching movies, which happens to be the overwhelming favorite leisure activity onboard nuclear submarines. Given the role of recreation in maintaining psychological homeostasis during periods of stress, it is highly likely the crew will have access to extensive libraries that will include literature, prerecorded programming, feature films, educational materials, and special broadcasts.

One recreational activity that may be implemented en route to Mars is the weekly lecture series, a tradition established by polar explorers. The subjects of these lectures could include topics of interest such as Martian geography, the finer points of aerocapture, and instruction in assisting the crew medical officer (CMO) in the event of a crewmember suffering a rapid decompression event.

In-flight medical care

A manned Mars mission will be one of great medical significance and it is probable at least one of the crew will be a physician with surgical training, and the other crewmembers will have received extensive CMO training. This training will be necessary since the weight and volume restrictions on the vehicle will severely limit the availability of surgical and anesthetic equipment to cover all but the most likely situations.

Medical care will also extend to ensuring the behavioral health of crewmembers, a requirement involving a team of NASA psychiatrists and psychologists with extensive experience working with astronauts and their families during missions. Much of the monitoring of astronauts' behavioral health will be by means of private psychological conferences (PPC) and CMO-administered psychological diagnostic tests. The results of the diagnostic tests will be sent to the crew surgeon at Mission Control for review and any recommended therapeutic response. Some might question the need to monitor the crew so closely, arguing that the crew selected to travel to Mars will most likely be the most thoroughly scrutinized and rigorously screened humans in history. However, while a serious or mission-threatening psychological incident may be unlikely, the consequences of a severe psychosis would be catastrophic for an expedition with no abort capability.

Crewed initial surface operations

Once the crew has survived the hazards of Mars orbit insertion (MOI), atmospheric deceleration, aerocapture, and entry, descent, and landing (EDL), they will commence surface operations.

Immediately after landing, the Commander will run through the post-touchdown checklist and a decision to stay/no stay will be made based upon the integrity of the landing site and any damage to the vehicle or its systems. The crew will then prepare for a period of adaptation to Martian gravity and their first steps on the Martian surface. To bodies adapted to life in zero-gravity, becoming accustomed to walking around the vehicle will be disorienting, but, after 3 or 4 days, the crew will prepare for the first Martian EVA. The most articulate member of the crew will egress first and, holding tightly onto the handrails of the ladder, make his/her way slowly to the surface.

Crewed long-term surface operations

After the daily ritual of showering and eating a breakfast of reconstituted cereal and Starbucks coffee, the crew will prepare for planned surface activities. After donning spacesuits, two crewmembers will enter the egress chamber and pass through the airlock, before emerging onto the surface of Mars, while the third crewmember remains inside, preparing to support and coordinate surface activities (Table 6.3).

One of the first tasks will be to unstow the drill rig and begin the search for water and evidence of life. By extracting core samples (Figure 6.12) from various sites over a period of several weeks, the crew will characterize the local area, before moving farther afield using the unpressurized rover. After 6–8 hr performing their surface activities, the astronauts will head home to the habitat. Back inside, the astronauts will relax in the small but comfortable confines of their habitat, helping each other prepare the evening meal.

Table 6.3. Scientific and human interest-driven activities.

Scientific activities			
Activity	*Indoor*	*Outdoor*	*Robotic*
Geology	Rock analysis Geochemistry Sample storage Age dating Teleoperate rovers	Field geology Mapping Geomorphology Stratigraphy Drilling	Sample collection Aerial reconnaissance Local resolution maps Multispectral mapping
Geophysics	Displays Data analysis System operations	Active seismic electromagnetic sounding	Local regional geophysical network (seismic)
Climate	Evolved gas analyzer	Hydrologic history Recent cyclic changes	Aerial reconnaissance
Meteorology	Display Atmosphere composition	Outpost meteorology station Tethered balloon	Regional network
Exobiology	Culture samples Planetary quarantine Back-contamination controls	Explore promising environments Hydrothermal areas Deep subsurface Drilling	Robotic field work

Human interest-driven activities	
Arrival on Mars	Crew connects power, assembles habitat, and raises flag(s)
Outpost set-up	Crew assembles structures, power systems, deploys radiators
Health maintenance	Crew under routine medical surveillance requiring tests for radiation exposure, exercise capacity, and bone loss
Crew meetings	Daily discussions concerning recent and future exploration activities. Changes to exploration plan discussed
Crew recreation	Sightseeing excursions on foot and by rover

Departure preparations and departure

During their final days on Mars, the crew will conduct pre-departure operations, including cleaning up the habitat, disposing of trash, placing systems in standby or off mode, and verifying ascent vehicle systems. On the final day on Mars, the crew will conduct a final EVA to load samples into the cargo bay, and then prepare for return to Martian orbit.

Figure 6.12. Artist's rendering of a Mars crew extracting core samples. Image courtesy: NASA.

Having described the types of missions an astronaut may be selected for in the next two decades, it is also worthwhile considering some of the training common to these missions. Due to complexities of the immediate and near-future expeditionary missions described here, it is perhaps virtual environment and analog training that will become increasingly relevant features of preparation.

ANALOG AND VIRTUAL ENVIRONMENT TRAINING

Virtual Environment Generator training

Prior to their first ISS mission, astronauts receive some of their familiarization training by using a Virtual Environment Generator (VEG). The VEG[6, 7] is a virtual reality (VR) system that can simulate certain aspects of microgravity, assist in navigating new environments, such as the ISS or lunar habitat, and serve as a countermeasure to spatial disorientation. The VEG (Figure 6.13) comprises a head-mounted display (HMD), the position and orientation of which command a computer to generate a scene corresponding to the position and orientation of the operator's head. This synthetic presence permits the operator to move around in the artificial world of the ISS, or even traverse the Martian surface.

When astronauts don the VEG equipment, they are presented with an image of the space station's/habitat's interior and a space-stabilized virtual control panel with an image of the astronaut's hand in the HMD. As the astronauts move their hands,

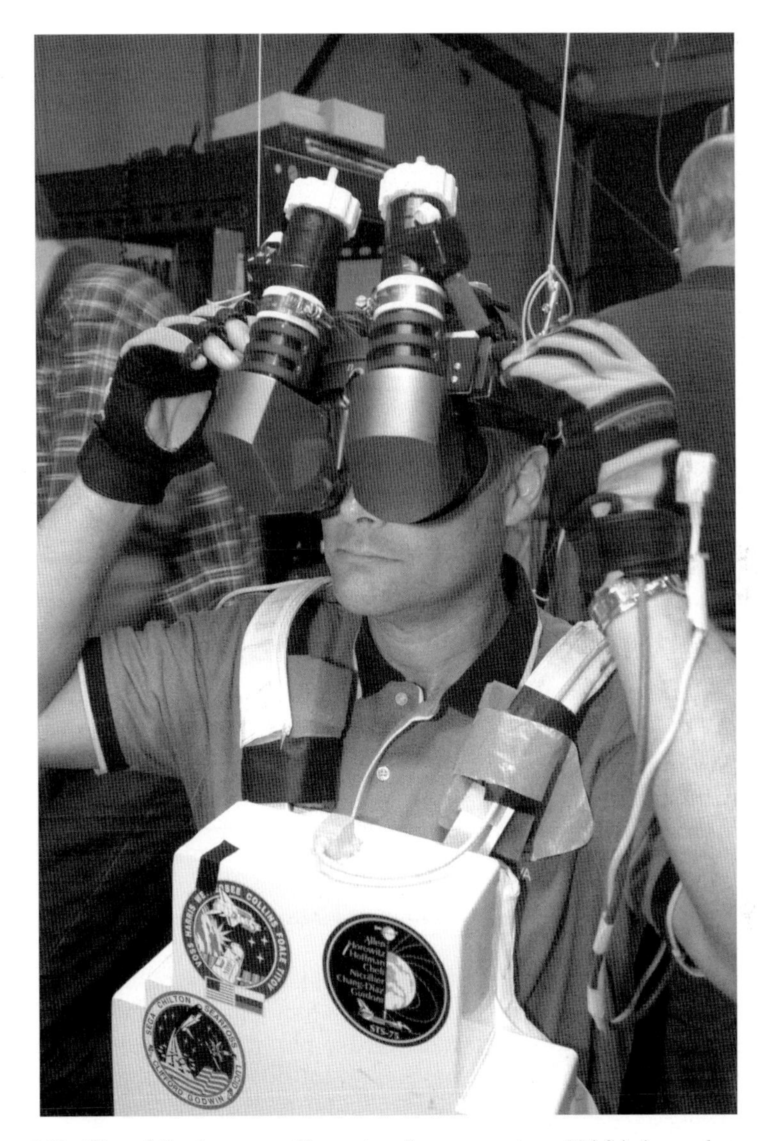

Figure 6.13. Virtual Environment Generator. Image courtesy: NASA (*see colour section*).

the virtual hand also moves. Collision detection software in the graphics computer detects when the operator's hand penetrates the virtual control panel, enabling the astronaut to interact with the virtual switches or objects to control events within the habitat. Astronauts are also able to manipulate objects in the virtual habitat and to experience resistance to movement, texture, mass, and compressibility, thanks to the haptic (tactile) and force feedback systems. To help astronauts in the virtual habitat, the system has been designed to provide auditory cues when an object is grasped or dropped, or when a virtual switch is operated. This synthesis of visual and auditory cues augments the visual information presented to the operator, thereby enhancing the performance of crewmembers within the habitat.

Data compression techniques result in the virtual habitat containing all objects one expects to see in the real habitat. Software also takes into account the effect of human behavior and the effect of collision for real-time operation, which means no matter how fast the operator moves through the environment, he/she experiences no visual lags. The real-time operation results in the operator experiencing the high degree of realism and interactivity necessary to allow crewmembers to perform tasks necessary for training.

Analogs

Throughout the history of space exploration, analog environments have played a vital role in astronaut training. Given the potential of high-fidelity simulations to prepare astronauts to conduct exploration and science activities, it is inevitable crewmembers bound for the Moon and Mars will spend a significant part of their training in one or more of the analog environments described here.

Antarctica

Antarctica has a number of characteristics making it a serious candidate for use as a planetary simulation facility. First, the continent is remote and hostile, as evidenced by the fact the United States Antarctic program has experienced more than 60 deaths since 1946. However, with proper logistical support and safety awareness, training can be conducted with acceptable levels of risk. Second, the isolation of the continent serves as an ideal analog for studying human factors and small human populations over long periods of time. During the winter months, those living on the continent are completely shut off from the rest of the world except for radio communications and the occasional airdrop resupply. What better place to simulate a lunar or Martian mission? Third, the Antarctic Treaty provides a proven framework for international cooperative exploration and scientific efforts similar to the model envisaged for a human mission to the Moon and Mars.

Haughton Mars Project

While not as isolated as Antarctica, the polar region of Canada has played host to a planetary simulation facility for more than a decade. Designed to serve as an analog for Mars, the Haughton Mars Project (HMP), supported by both the American and Canadian space agencies, is perhaps the most successful and widely known planetary analog.

The HMP is an international interdisciplinary field research project conducted in the vicinity of the Haughton impact crater on Canada's Devon Island. The project, conducted jointly by the Mars Institute, NASA, Search for Extraterrestrial Intelligence (SETI), and the Canadian Space Agency (CSA), utilizes the Mars-like features of the island and the crater to develop and evaluate new technologies and field operating procedures that may one day be used on a manned mission to Mars.

Crewmembers

Each summer, dozens of researchers, students, support staff, and media journey to the HMP site. Some spend the entire summer there, while others rotate in and out for shorter stays of a few days or couple of weeks. In 2008, HMP celebrated its 12th field season, featuring scientists and researchers from Simon Fraser University (SFU) in Vancouver, NASA's Ames Research Center (ARC), the CSA, the Mars Institute, Hamilton Sundstrand, McGill University, and the National Space Biomedical Research Institute (NSBRI). Most crewmembers arrive courtesy of Marine C-130 crews, who support HMP with the delivery of expeditionary equipment, research gear, and exploration vehicles. Getting around the HMP site is either by all-terrain vehicle or simply by walking, often while wearing a spacesuit similar to the one that may be used on the Martian surface.

Environment

Although the Haughton Crater is a cold, dry, barren, dusty, and windy place with an average temperature of only –17°C, in comparison with Mars, which has an average temperature of –60°C and is drenched in lethal radiation, the HMP analog is a comparatively pleasant place. However, while elements of the Martian surface environment may be absent, the crater is a step in the right direction to evaluating strategies that will help the future exploration of Mars.

The Haughton Crater is characterized by ancient lakebeds, rocky terrain, occasional valleys, and steep-walled canyons. Although rocks on Mars may have a different composition, the physical properties of the impact deposits still provide a valuable analog for the scientists stationed there. Recent neutron spectrometry data from the Mars Odyssey spacecraft have provided evidence of ice deposits at shallow depths in the Martian subsurface, similar to subsurface ice at the Haughton Crater.

In addition to these subsurface ice deposits, the Haughton Crater provides remnant signatures of hydrothermal activity and networks of channels bearing a close resemblance to the small valley networks on Mars. Perhaps the most valuable attribute of the Haughton Crater is the presence of life. Despite the high ultraviolet radiation environment during the summer and the harsh weather during the winter, Haughton Crater microorganisms are able to survive and adapt – a capability with potential implications for the search for life on Mars.

Figure 6.14. NASA FS, Josef F. Schmid (left), and astronaut, Jose M. Hernandez, participate in an underwater extravehicular activity session during the 12th NEEMO mission. Image courtesy: NASA (*see colour section*).

NASA Extreme Environments Mission Operations Project

Aquarius is a unique space analog research station (Figure 6.14) anchored 16 km from Key Largo in the Florida Keys National Marine Sanctuary. The facility, the only undersea laboratory in the world, provides living quarters permitting resident astronauts (aquanauts) to stay on the seafloor for extended periods. The air inside Aquarius is pressurized to counterbalance the weight of surrounding water, which means crewmembers must breathe pressurized air. Breathing pressurized air for several days means the aquanauts are saturated with nitrogen, requiring them to perform a lengthy 16-hr decompression to the surface at the end of their 2-week stay.

Undersea missions

The Aquarius environment closely resembles the living and working conditions astronauts face in space so it is not surprising NASA has used the facility to conduct 2-week missions and extravehicular activity (EVA) simulations (see Chapter 7).

These simulations are an element of the NASA Extreme Environment Mission Operations Project (NEEMO). The 2-week missions are planned in a very similar manner to phases of an ISS increment, with each crewmember being assigned a detailed timeline of activities and specific timeslots for each activity, including outreach, sleep, chores, and hygiene. One of the many activities performed by NEEMO's aquanauts are dives, planned analogous to space EVAs, requiring advanced planning and a series of objectives, timelines, and goals. While conducting underwater EVAs, astronauts practice communication tasks such as ship-to-ship calls to the ISS, and even construction assignments such as building solar arrays.

Exploration operations

During their 2-week undersea increments, crewmembers have the opportunity to practice all sorts of operations and techniques that may one day be conducted for real as part of a manned Moon/Mars mission. For example, astronauts practice operating remotely operated vehicles (ROVs), providing training similar to that required for operating rovers on the lunar surface. They also go on occasional treasure hunts, trying to figure out the best way to perform search and rescue. The treasure hunts begin with a crewmember dropping markers at random locations along the reef. Using Doppler navigation and transponders, crewmembers then go outside the habitat and try to find the markers using various search techniques.

While the Aquarius facility is undoubtedly a high-fidelity simulation imitating many aspects of spaceflight, due to the 2-week duration, astronauts don't have an opportunity to experience the isolation and confinement awaiting them during a 6-month stay onboard the ISS or a 3-year journey to Mars. For this, an isolation chamber is required.

MARS500

Recently, ESA prepared to simulate a 500-day mission to Mars, dubbed Mars500. The aim of the first part of the study was to seal four carefully selected candidates inside an isolation chamber for 105 days, that commenced March 31st, 2009. This period was then followed by the full isolation period of 520 days. One section of the isolation chamber (Figure 6.15) used in the study simulated the spacecraft that would transport them to and from Mars, and another section simulated the landing module that would transfer them to and from the surface of Mars.

The experimental facility (Table 6.4), which includes the isolation facility, operations room, and technical facilities, is located in a building of the IBMP in Moscow. The layout comprises four hermetically sealed interconnected habitat modules and one external module, which will be utilized as an analog for a Martian surface excursion. The volume of all habitat modules is 550 cubic meters.

Once sealed inside the isolation chamber, the participants were subject to the same restrictions as an astronaut embarking upon a space mission. For example, voice

Figure 6.15. Overhead view of the isolation chamber used in the ESA's Mars500 mission. Image courtesy: ESA.

contact with family and friends was conducted via a simulated Mission Control center and, as with all astronauts, they were subject to extensive testing and evaluation. Scientific investigations included analysis of urine, blood, ECG, sleep quality, and the influence of exercise and food supplementation, in an attempt to aid the development of potential countermeasure tools and techniques that may be implemented during the real manned mission to Mars.

Due to training demands, it is unlikely astronauts will be called upon to spend 500 days in an isolation chamber! However, given the unique psychosocial demands associated with an expedition to Mars, it is probable mission planners will require a Mars crew to spend at least a couple of months in a similar facility. Such an exercise would satisfy mission planners and crew that no interpersonal conflicts exist that might jeopardize the mission, and also serve to increase group cohesiveness in a confined environment.

This chapter has described the types of missions astronauts selected in 2009 may be assigned to in the next 20 years. It has also described some of the specialized training that will likely feature prominently in the preparation of astronauts bound for the lunar and Martian surfaces. However, for the next 10 years at least, astronauts will be confined to orbits of the Earth onboard the ISS. Preparing for these missions requires astronauts to spend 18 months in what is known as Increment-Specific Training.

Table 6.4. Mars500 experimental facility.

Module	*Dimensions (m)*	*Designation*	*Description*
1	3.2 × 11.9	Technical–medical	Houses two medical berths, a toilet, and equipment for routine medical examinations. Also includes equipment for performing tele medicine and diagnostic investigations
2	3.6 × 20	Living quarters	The main living quarters comprise six individual compartments, including a kitchen–dining room, main control room, and lavatory. The 2.8 × 3.2 m crew compartments each have a bed, desk, chair, and shelves
3	6.3 × 6.17	Mars landing module	This compartment will be used only during the 30-day Mars orbiting phase. It comprises three bunk beds, two workstations, a lavatory, control and data collection system, communications system, ventilation system, waste treatment facility, and a fire suppression system
4	3.9 × 24	Storage module	Comprising four compartments Compartment 1 Fridge Compartment 2 Storage of non-perishable food Compartment 3 Location of experimental greenhouse Compartment 4 Bathroom, sauna, and gym

REFERENCES

1. Bufkin, A.; Tri, T.O.; Trevino, R.C. EVA Concerns for Future Lunar Base. *Second Conference on Lunar Bases and Space Activities of the 21st Century*, Paper No. LBS-88-214. Lunar and Planetary Institute, Houston, TX (1988).
2. Simonds, C.H. Sintering and Hot Pressing of Fra Mauro Composition Glass and the Formation of Lunar Breccias. *American Journal of Science*, **273**, 428–439 (1973).
3. Christiansen, E.L.; Euker, H.; Maples, K.; Simonds, C.H.; Zimprich, S.; Dowman, M.W.; Stovall, M. *Conceptual Design of a Lunar Oxygen Pilot Plant*, EEI Report No. 88–182. Eagle Engineering, Houston, TX.
4. Doll, R.E.; Gunderson, E.K.E. *Hobby Interest and Leisure Activity Behaviour among Station Members in Antarctica*. San Diego, California. US Navy Medical Neuropsychiatric Research Unit. Unit Report No. 69–34 (1969).
5. Eberhard, J.W. *The Problem of Off-Duty Time in Long-Duration Space Missions*,

three volumes, NASA CR 96721. Serendipity Associates, McLean, Virginia (1967).

6. Cater, J.P.; Huffman, S.D. Use of Remote Access Virtual Environment Network (RAVEN) for Coordinated IVA-EVA Astronaut Training and Evaluation. *Presence: Teleoperators and Virtual Environments*, **4**(2), 103–109 (Spring 1995).
7. Chung, J.; Harris, M.; Brooks, F.; Kelly, M.T.; Hughes, J.W.; Ouh-young, M.; Cheung, C.; Holloway, R.L.; Pique, M. Conference Proceedings: *Exploring Virtual Worlds with Head-Mounted Displays, Non-Holographic Three-Dimensional Display Technologies*, Los Angeles, January 15–20, 1989.

7

Mission training

After spending years performing various technical assignments supporting astronauts who are already in space, and those training to go, ascans are finally called in to meet the Chief of the Astronaut Office. The Chief Astronaut, the most senior astronaut position, serves as head of the NASA Astronaut Corps and is the principal advisor to the NASA Administrator on the subject of astronaut training and operations. A call to his/her office usually means one thing and one thing only: mission assignment.

The meeting usually begins with the Chief Astronaut making complimentary remarks about how well the ascan did in his/her technical assignments and ends with asking whether the ascan would still like to go to space. Most ascans don't take too long answering that question!

For the foreseeable future, spaceflights conducted by American, Canadian, and European astronauts will be to the International Space Station (ISS). At the time of writing, an Expedition crewmember spends almost 1,800 hr training for an ISS (Figure 7.1) mission. This time includes approximately 330 hr learning about US ISS systems, 350 hr learning about Russian ISS systems and the Soyuz spacecraft, 400 hr learning to do spacewalks (300 in US spacesuits and 100 in Russian spacesuits), 65 hr of medical training, 150 hr of science experiment training, 300 hr of language training, and 200 hr of robotic arm training. The training is conducted at dozens of locations around the world, some of which are listed in Table 7.1.

Table 7.1. Mission training centers.

Center	*Description*
Johnson Space Center (JSC), Houston, Texas	JSC is home base for NASA's astronauts and a home away from home for visiting cosmonauts and Expedition crewmembers from other countries. JSC is also the primary training location for Expedition crews. Teams of professional instructors use its classrooms, standalone training facilities, integrated simulation environments, and laboratories to help crewmembers to prepare for their missions

Table 7.1. *cont.*

Center	*Description*
Kennedy Space Center (KSC), Florida	KSC is located on Florida's Atlantic Coast. NASA's new family of launch vehicles, *Ares I* and *Ares V*, will be launched from KSC starting in 2015 and will eventually carry crews and cargo to the Moon. Expedition crews also visit KSC to train for launch and practice emergency procedures
Canadian Space Agency (CSA), Quebec, Canada	Canada contributes essential components of the ISS, such as the Mobile Servicing System, which includes Canadarm2 and the Mobile Base System. In Montreal, astronauts receive robotics training to prepare them for complex arm operations. They also use the Virtual Operations Training Environment located at the CSA's headquarters in Saint-Hubert, Quebec. It provides crew members with an immersive virtual-reality environment in which they can watch a simulated arm move in three dimensions and gain a deeper understanding of the Canadarm2's movements relative to external structures on the ISS
Gagarin Cosmonaut Training Center (GCTC), Star City, Russia	The GCTC, located near Moscow, is the primary training facility for Russian elements of the ISS. Instructors use classrooms, simulators, and full-scale mockups to provide crewmembers with the knowledge they need to work in the Zvezda Service Module and the Zarya Module. Also located at GCTC is the Hydrolab, Russia's equivalent of NASA's Neutral Buoyancy Lab at JSC, which provides a realistic training environment for spacewalks performed out of the Russian airlock in Russian spacewalking suits
Baikonur Cosmodrome, Baikonur, Kazakhstan	Baikonur has been home to Russian space launches since Kazakhstan was a part of the old Soviet Union. The complex of launch pads and support facilities are known collectively as the Baikonur Cosmodrome. Expedition crews and taxi crews who will travel to the ISS aboard a Soyuz spacecraft visit Baikonur for some of their Soyuz training
European Astronaut Centre (EAC)	Provides crew training for the *Columbus* module, including four ESA payload racks (European Drawer Rack, Fluid Science Laboratory, European Physiology Module, and the Biological Laboratory) and the Automated Transfer Vehicle (ATV). Also houses ESA's Neutral Buoyancy Facility

Figure 7.1. Current configuration of the International Space Station. Image courtesy: NASA.

INCREMENT-SPECIFIC TRAINING

As soon as a NASA, European Space Agency (ESA), or Canadian Space Agency (CSA) astronaut is assigned to a mission, Increment-Specific Training (IST) begins. The word *increment* is a favorite term among space agencies and simply refers to the time period between crew exchanges onboard the ISS. Sometimes, IST is also referred to as mission-specific training. During the 18-month IST phase, the prime and backup crews train together and learn everything they need to know for their mission. The team training is a vital element of the preparation, since crewmembers must become familiar with each other before spending the better part of half a year cooped up in an enclosed environment no larger than a three-bedroom house. Also, by spending so much time together during training, the crewmembers learn how to work effectively and how to assign roles and responsibilities among themselves.

Crew qualification levels

During IST, astronauts attain three crew qualification levels: user, operator, and/or specialist level. For each onboard system, a set of minimum qualifications needed to safely operate and maintain the system is pre-defined. For example, there is normally one specialist, one operator, and one user. Each crewmember, while being specialist

Figure 7.2. Canadian astronaut, Robert Thirsk, Expedition 20/21 flight engineer, participates in a training session in an International Space Station mockup/trainer in the Space Vehicle Mock-up Facility at NASA's Johnson Space Center. Image courtesy: NASA.

for some systems, will be an operator or only a user for other systems. Consequently, the training program for each crewmember is individually tailored to his or her set of tasks and existing qualification levels.

Emergency training and supplementary training

IST also includes training to deal with what space agencies euphemistically refer to as off-nominal situations (what most people would call an emergency!), failure analysis, and recovery/repair activities. Astronauts also learn how to run the science experiments scheduled for their mission, and are trained in the delicate art of rendezvous and docking spacecraft with the ISS, such as ESA's Automated Transfer Vehicle (ATV). The IST phase also includes multi-segment training, implemented at NASA's Space Station Training Facility (SSTF), located at Johnson Space Center (JSC). The SSTF (Figure 7.2) is a replica of the entire ISS in which the astronauts can train for emergency situations that may affect the whole station.

Although astronauts, whether they belong to NASA, the CSA, or the ESA, follow a very similar training schedule, there are some differences. These differences are usually related to systems onboard the ISS that have been developed by the agency to which the astronaut belongs. For example, the CSA's most prominent contribution to the ISS is the robotic arm, so when Canadian astronauts are assigned to a mission, they are usually designated as the robotic arm operator. The ESA's most high-profile contributions include the *Columbus* module and the ATV, so it is not surprising ESA astronauts receive training specific to these. While most of this chapter is devoted to NASA's astronaut IST schedule, it is worthwhile recognizing some of the training offered by other ISS partner agencies, such as the ESA.

EUROPEAN SPACE AGENCY TRAINING

Much of the IST unique to the ESA focuses on the ESA's 20-tonne ATV (about the size of a London double-decker bus), which is the largest spacecraft ever built in Europe. The unmanned vehicle is used to ferry cargo to the ISS and to raise its orbit. To learn how to operate the ATV (Figure 7.3), astronauts undergo extensive training at the ESA's European Astronaut Centre (EAC), in Cologne, Germany. The training is divided into two 1-week modules. In addition to receiving theoretical instruction on ATV missions and malfunction scenarios, the astronauts also spend time inside the high-fidelity ATV mockup and ATV simulator. The mockup helps the astronauts orientate themselves to the various parts of the ATV, while the simulator teaches the astronauts how to control the software that navigates the spacecraft. Helping the astronauts during their instruction are seven instructors, who teach the following four phases of ATV training (Table 7.2):

1. Core lessons
2. Rendezvous and docking
3. Attached phase operations
4. Emergency training

Table 7.2. The European Space Agency's ATV increment-specific training flow.

	ISS pre-requisite training	*ATV Training Period 1*	*ATV Training Period 2*
NASA	Inventory Management System		
ESA		*CORE Block* Vehicle systems & Ops concept *RVD Block* RVD & undocking Emergency procedures Ingress/egress ops	*Attached Phase Ops* Cargo transfer, waste disposal, departure ops Simulations: RVD undocking & departure
GCTC	Emergency equipment. Russian Docking System	Russian Segment Systems relevant for ATV ops	RVD, undocking and departure proficiency sims

RVD: Rendezvous and Docking

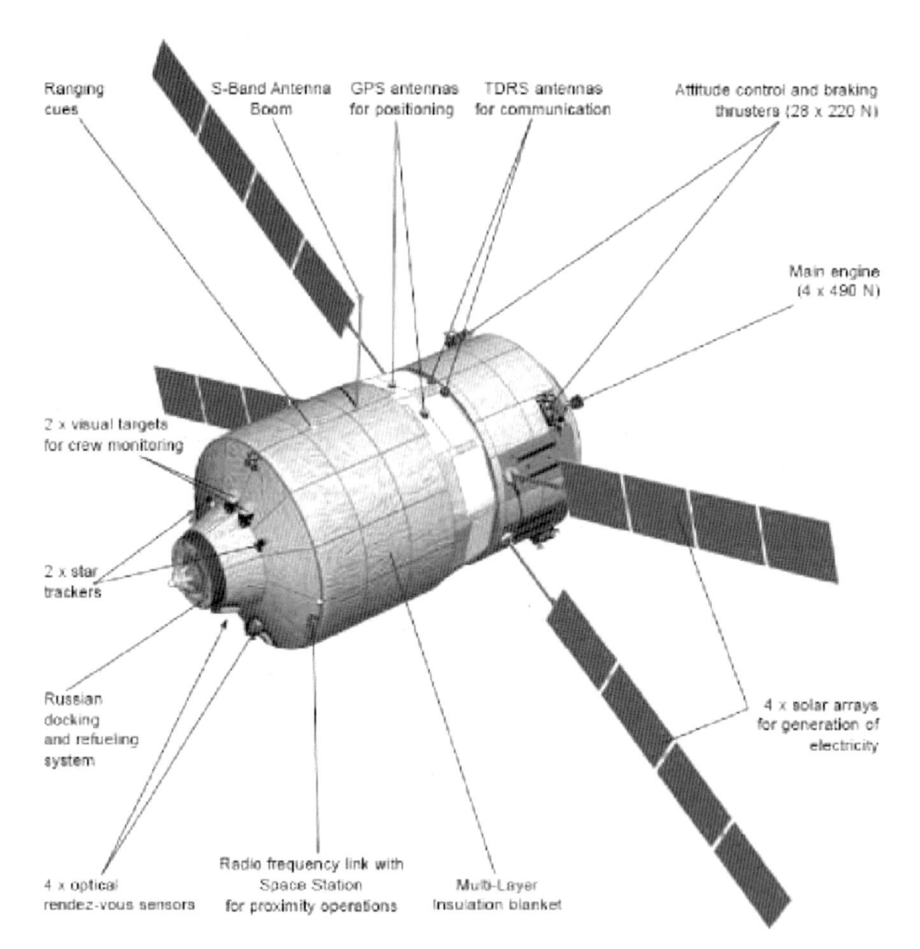

Figure 7.3. The European Space Agency's Automated Transfer Vehicle. Image courtesy: ESA.

Core lessons

Before astronauts commence ATV training, they must meet certain pre-requisites. For example, they must have experience with the Russian Docking System (RDS), needed because the ATV docks with the Russian Service Module Zvezda. The astronauts also must have experience operating the American Inventory Management System (IMS), a barcode reader system similar to that used in supermarkets, which helps the astronauts find items on the ISS and keep it tidy (Panel 7.1).

Using the reader, astronauts scan the cargo "out" of its location inside the ATV and "in" to the new location where it is stored in the ISS. Using this method, a computer can track the inventory onboard the ISS and the astronauts do not have to memorize the whereabouts of every single item. Although time-consuming, the

Panel 7.1. Keeping a space station tidy

Asset tracking for space exploration relies on a barcode-based system interfacing with an Inventory Management System (IMS) database. Unfortunately, for the astronauts, using this system for inventory management is very labor-intensive and occasionally an update is missed and an item's status becomes unknown. In fact, on any given day onboard the ISS, 3% of US items in the IMS database are listed as "lost", meaning the item was not in the location it was showing in the IMS the last time the crew looked for it! This obviously creates problems for Mission Control, since, if a critical item is lost, Mission Control must decide whether to allocate critical crew time to continue looking for the item. Alternatively, if there is a spare on the ground, Mission Control must decide whether to launch a replacement on an upcoming mission. Either option is very costly. To keep track of items onboard, an ISS crew is allotted 20 min every day to make updates to the IMS. In reality, a much larger portion of their day is dedicated to this task.

system is invaluable, since there is a crew rotation every 6 months. Once they are proficient in the operation of the IMS, astronauts begin the theoretical instruction that provides them with a general overview of what the ATV is, how it works, and what happens during an ATV mission. These lessons are followed by a tour through the ATV mockup where the astronauts put into practice what they learnt in class. After the mockup tour, the astronauts prepare for the rendezvous and docking phase. But, before this, they must learn about the ATV's systems and its navigation sensors.

Rendezvous and docking

The second set of ATV lessons focus on rendezvous and docking maneuvers. In this series of lessons, the astronauts are taught how to perform the operations needed when the ATV arrives at the ISS. This phase of instruction begins with an introduction to the ATV's concept of operations.

After launch, the ATV may orbit the Earth for a number of days before commencing rendezvous with the ISS. To test all systems are functioning nominally, the ATV Control Centre (ATV-CC), in Toulouse, France, and the ISS crew send commands to the spacecraft and check the ATV responds correctly to each command. These maneuvers are intended to demonstrate satisfactorily that the sensors and the guidance, navigation, and control system onboard the ATV can safely approach and dock with the ISS. Once this is accomplished, ATV docking is guided by sensors that recognize the correct docking area and initiate the process in a very precise manner. Since the ATV docking is automated, you may wonder why astronauts need so much training. One reason is the ATV is a new vehicle with little inflight experience. A second reason is safety. Mission managers haven't forgotten the collision of a Russian Progress spacecraft with the Russian space station Mir on June 25th, 1997. On that fateful day, Mir commander, Tsibliev, was remotely commanding the approach of Progress to Mir's *Kvant* module. This required guiding the Progress using a television monitor. The Progress was difficult to see against the cloudy Earth background at the time of the attempted docking, with the result that it went off course and collided with a solar array on the *Spektr* module and then the module itself. A large hole was made in a solar panel, one of the radiators was buckled, a hole was punched into *Spektr*'s hull, the module began to depressurize, and the crew nearly had to abandon the station. Needless to say, the lesson that accidents can happen even with a trained astronaut performing the rendezvous wasn't lost on mission planners, which is why the "automated" ATV always has an extra pair of eyes!

While onboard the ISS, the crew can see the ATV using a video camera mounted on the ISS and a specially designed target on the ATV. With this video image, the crew can judge whether the ATV is flying within a corridor aligned with the docking port on the ISS. This ability enables the crew to monitor the ATV and provides an additional level of safety. Finally, given the ATV's size, any docking maneuver with the ISS is considered a risky operation and every possible precaution is used. To train for this docking and undocking maneuver on the ground, astronauts are trained in the simulator and in the high-fidelity ATV mockup. In the simulator, the crew learns how to handle the Tele-operator Control System (TORU) control workstation. The TORU includes the Simvol screen, which displays an overlay of rendezvous data, enabling the astronauts to monitor the ATV's approach to the ISS using no more than a Russian laptop. The second step of the docking and departure training is conducted inside the mockup, where the astronauts operate in situ. There, an instructor guides the astronauts while simultaneously acting as the ground voice communication. Once they have been trained in the basics of rendezvous and docking, the astronauts move on to attached phase operations.

Attached phase operations

In the mockup, the astronauts conduct operations that are necessary once the ATV is attached to the ISS. For example, they learn how to open the ATV's hatch, an operation known as "ingress", and are taught how to connect the complex water transfer system used to pump water from the ATV tanks to the ISS tanks. They also learn how to transfer gases such as oxygen and nitrogen from the ATV to the ISS. Another important component of the ATV attached phase training is spent learning to prepare the ATV for departure. Following departure from the ISS, the ATV is simply de-orbited and allowed to burn up in the Earth's atmosphere. Preparing to conduct such a maneuver may seem relatively simple, but, in low Earth orbit (LEO), even getting rid of waste is a complicated business, since the ATV must first be loaded to ensure the weight of the cargo is correctly balanced. If the ATV is off-balance, the thruster control of the vehicle is difficult to calculate and there is a risk the vehicle will become difficult to control.

Emergency operations

Emergency operations training comprises a series of simulations conducted inside the ATV mockup. The training focuses on two situations: fire and depressurization. By conducting the simulations, the crew gradually improves its ability to react to these two critical emergencies in a timely manner.

Columbus training

In addition to being trained in ATV operations, resident ISS crews must also be trained in the nominal operation and malfunction of systems onboard the ESA's *Columbus* laboratory. The 4.5-m diameter *Columbus* laboratory (Figure 7.4) is the ESA's biggest single contribution to the ISS. Equipped with research facilities offering extensive science capabilities, the laboratory enables Earth-based researchers, together with the ISS crew, to conduct experiments in the disciplines of life sciences, materials science, and fluid physics. During the advanced phase of training, astronauts will have received introductory lessons on *Columbus*'s five systems: Data

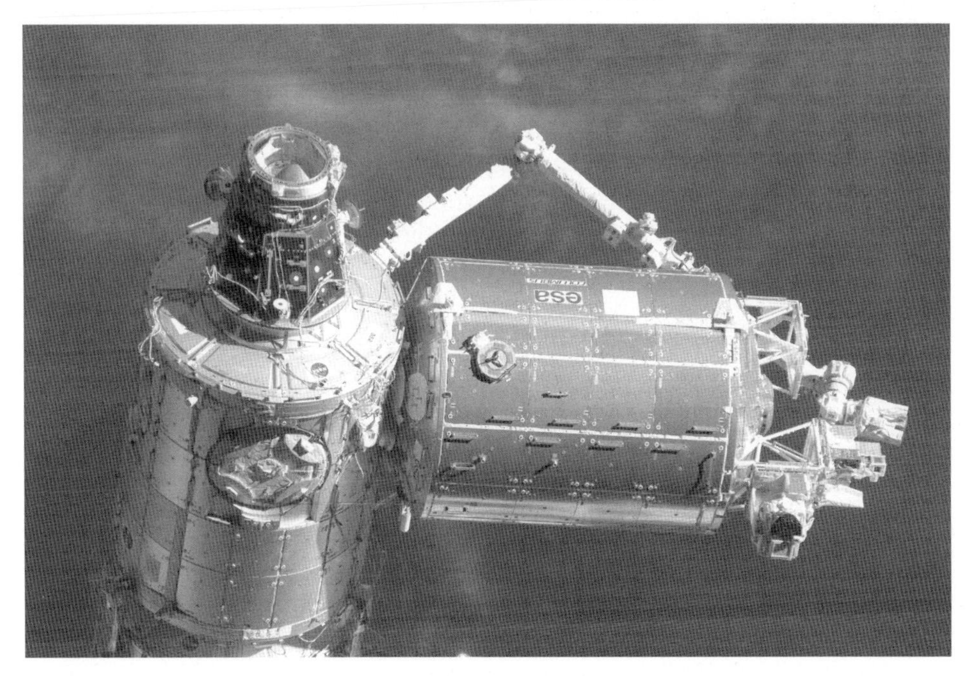

Figure 7.4. The European Space Agency's *Columbus* laboratory. Image courtesy: ESA.

Figure 7.5. The *Columbus* laboratory mockup contains all the racks and experiment facilities currently mounted in the flight version docked with the ISS. Image courtesy: ESA (*see colour section*).

Management System (DMS), Electrical Power Distribution System (EPDS), Thermal Control System (TCS), Environmental Control and Life Support System (ECLSS), and the Communication System. During the IST phase, the astronauts spend more time learning malfunction operations as well as preventive and corrective maintenance. This training is delivered in three visits to EAC, during which operators and specialists receive 84 and 104 hr of training, respectively.

In common with the ATV training, a number of well equipped mockups and simulators are available at EAC to provide astronauts with high-fidelity facilities. In addition to the computer-based training facilities, EAC houses the *Columbus* mockup (Figure 7.5) and the Columbus Trainer Europe, which houses a simulator permitting flight-like telemetry to the crew and ground support personnel.

Once astronauts have completed the ATV and *Columbus* training phases, they integrate back into the regular IST schedule, which may mean training with their NASA and CSA colleagues in Houston, Star City, and Japan. Since NASA is responsible for delivering the lion's share of IST preparation, the rest of this chapter is devoted to describing a typical 18-month IST period.

EXPEDITION TRAINING

For NASA and CSA astronauts, IST/Expedition training begins at JSC in Clear Lake, near Houston, Texas. During their first training class, the astronauts learn how to do the things they'll do every day aboard the ISS. They learn how to read procedures written by people on the ground, which they must complete correctly in orbit. They also learn how to use the ISS's laptop computers and the software. They spend time in simulators, where they work in a mockup of the ISS and practice what they have learned so far. At the beginning of each new training session, the ISS training lead and the training team give each astronaut a Crew Training Notebook (CTN). This binder contains all the handouts and study aids the astronauts need during a particular training session. This introductory phase of training is then followed by one of many trips to Russia.

Once an astronaut is assigned to an ISS flight, they require about 18 months of training. However, it may take longer than 18 months if the crewmember hasn't received specific training on certain ISS systems, or if they don't have the requisite language skills. It also helps if the astronaut has trained as a backup to another expedition crewmember and has learned to speak Russian. Ideally, as an astronaut working on the ISS, it helps to have a good understanding of either French or German, given the truly international nature of the ISS program. During their 18 months of training, astronauts make several trips between NASA's JSC and the Gagarin Cosmonaut Training Center (GCTC) in Star City, Moscow, Russia. In addition to racking up airmiles between Houston and Moscow, astronauts also make as many as eight or nine overseas trips to other ISS member states. For example, astronauts tasked with operating DEXTRE* will visit Saint Hubert in Canada to train on the station's robotic arm. Some astronauts will also visit Japan to be trained on systems specific to Japan's *Kibo* module.

As soon as crewmembers are assigned to a mission, a Crew Qualifications and Responsibility Matrix (CQRM) is created. The CQRM document contains information about what each crewmember will be doing during their mission. Training staff in Russia and the US use the CQRM to decide whether a crewmember will be an operator or a specialist for ISS systems. Whereas an operator is only required to know how to operate a piece of equipment, such as sending a command to raise the temperature in one of the ISS modules, a specialist needs to know how to repair the system controlling the temperature inside the module.

Each ISS system, such as electrical, heating, cooling, and communications, has a separate training plan for operators and specialists, and crewmembers must know enough about every system to be at least an operator. Since being a specialist requires more training, astronauts are only trained to be a specialist on a few systems.

---

* Also known as the Special Purpose Dexterous Manipulator (SPDM).

Team assignment

The next step in the training plan is to assign the astronauts to a training team. The training team teaches the crew everything they need to know to have a safe and successful mission. In charge of the training team is the Station Training Lead (STL), typically a former instructor with many years' experience teaching astronauts (Panel 7.2). The team has one instructor for each of the eight main ISS systems, in addition to having instructors for the scientific experiments the astronauts will be conducting aboard the station, and other instructors to teach skills such as EVA.

Panel 7.2. International Space Station Training Lead

The STL has a team of instructors who help the astronauts during their mission training. Training begins in a classroom, where the STL's team provides students with a general overview of the ISS and its systems. From there, the crew moves to hands-on work in space station simulators, where they practice what they've learned. Once the crew has mastered the basics, the STL make things a little more difficult by preparing the crew for things that might go wrong. Basically, it's the job of the STL to try to find any shortcomings in the crew's ability. This contingency planning has been a vital part of all ISS-increment preparations because it is not enough for the crew to be prepared just for their primary mission objectives. They must also be trained for tasks from the preceding mission, in case something doesn't get finished, and in tasks for future missions, in case the current crew has time to get ahead.

The STL's team consists of experts in the various ISS systems, such as life support and robotics. Team members usually work in a specific training area before joining the integrated station training team. For each of the core ISS systems, there is someone who is an expert in that area. In addition to working directly with the astronauts, the STL and his/her team plans the crew's training, and tests the simulators the crew will use.

TRAINING FLOW FOR A GENERIC ISS MISSION

Russia Part I

An astronaut assigned to a mission may first travel to the GCTC in Star City. If they have worked as a backup (what NASA designates as ISS backup flow) for a previous mission, the chances are this will not be their first trip to Russia. The first trip may last between 4 and 12 weeks and will be the first of perhaps as many as six trips in 1 year. The first week is very much like attending university, with astronauts attending classes every day, 5 days a week. Each day typically comprises four periods of 1 hr and 50 min each with a lunch break between 1.00 and 2.00 in the afternoon. Classes usually begin at 9.00 in the morning and end just before 6.00 in the evening, so this aspect is a little different from being at university!

Living in Star City

Astronauts who train in Star City live in "The Cottages", which are a set of duplexes NASA built for their employees. The duplexes comprise six three-bedroom houses, and are situated a convenient 10-min walk from the training facilities. One of the cottages has a basement that houses a gym with weights and cardio machines, used by those training to fly to space or rehabilitating from a spaceflight. Another popular hangout is "Shep's Bar", located in the basement of one of the other cottages. The bar is named after Bill Shepherd (who built the bar), a NASA astronaut who spent so many years training in Star City that he decided it would be an idea to have a place where crews could congregate socially. Over the years, the bar has been used intensively, and donations from astronauts and other visitors have paid for a pool and ping pong table, a television, and stereo systems.

Shortly after arriving in Star City, astronauts are given a short tour of the training facilities, which usually comprise three or four buildings. Sometimes, depending on whether it is occupied, they have the opportunity to have a familiarization session in the Soyuz simulator, where they will spend many hours practicing drills and procedures. Once the tour is over, the astronauts start training.

Learning Russian

One of the first tasks for prospective ISS crewmembers is to learn Russian! This often proves a challenge for many astronauts, as their most recent memory of Russian was the introductory course they received while they were ascans! Astronauts must be able to understand Russian (Panel 7.3) so they can talk with Russian Mission Control Center and understand their Russian instructors who teach them how to operate systems on the Soyuz. Given the complexities of the language, all the astronauts have what they call "good Russian days" and "bad Russian days".

Praktika and konsultatziya

The first series of classes are mostly theoretical, although astronauts spend some time involved in *praktika* training on ISS mockup modules, such as the Service Module (SM) and the Functional Cargo Block (FGB). Eventually, each astronaut is required to enter the training mockups and demonstrate their knowledge to the instructors in what the Russians refer to as *konsultatziya* (exam). In the

Panel 7.3. Speaking Russian

Many astronauts agree that learning Russian is the hardest part of astronaut training. Not only must they master an alien alphabet, they must also train their ears and mouths to listen to and enunciate new sounds. If that wasn't enough of a challenge, astronauts must also grapple with a vocabulary rich with meanings. Protocols and agreements have to be translated from English to Russian, and then back into English to make sure translators have not skewed information.

For some of the most highly trained humans on (and off!) the planet, Russian language training can be a humbling experience. Some astronauts have described their first few Russian lessons as being similar to a first-grader attending graduate school. For most astronauts, it takes at least 6 months to achieve even a semblance of fluency. For others, it may take longer, which can be frustrating when one considers all oral exams and lectures in Star City are in Russian!

konsultatziya, the astronauts are quizzed on theoretical aspects of the module and must then demonstrate to the instructors how to operate every system in the module. Making the task even more demanding is the fact the *konsultatziya* are conducted in Russian, further compounded by the fast-talking Russian instructors! The exams are usually conducted in the same room used for lectures, so astronauts can touch or point to the hardware if they need to. For the test, the astronaut sits in the seat where their instructor usually sits, and two examiners plus the instructor sit opposite. Fortunately, for those astronauts who are not yet fluent in Russian, there is an interpreter in the back of the room!

In addition to being instructed on the SM and FGB, the astronauts spend much of their time learning about the Soyuz spacecraft, usually being taught by Russian Training Integrators (RTI) who are from the training department in Houston. The RTIs are experts on the Russian systems and make sure the astronauts receive exactly the right level of training. If they are not happy with how the Russian instructors teach a lesson, the RTIs will step in and cover it. The RTIs also ensure the astronauts have all the training materials in both Russian and English and, because they are experts on the systems, they help to prepare the astronauts for the exams.

Gradually, the astronauts work their way through the myriad Soyuz systems, ranging from life support to the functional idiosyncrasies of the spacecraft's toilet. After learning the theoretical aspects of the hardware, they have the opportunity to practice drills in the simulators. After several weeks of intensive studying, many welcome the opportunity to do something that takes place outside the classrooms.

Survival training

In between systems training and the challenges of struggling to learn what sometimes feels like an alien language, the astronauts have the opportunity to spend a few days in the wilderness as part of survival training, Russian style! For those astronauts whose last experience of survival training was as an ascan in the balmy climate of Rangeley, Maine, the Russian version is sometimes a shock to the system, especially those who are deployed to Russia during the winter!

For astronauts accustomed to Houston's hot and humid climate, survival training in a Russian winter definitely brings a touch of realism to the art of staying alive. With temperatures regularly below –20°C, survival training in the forests surrounding Star City takes on new meaning to North American astronauts, who usually use all their issued layers of arctic clothing to stay warm.

To simulate landing in a remote area, astronauts practice getting out of a capsule in all their survival gear before spending the night out in the open. Other survival exercises simulate landing in the water. In this exercise, astronauts must spend almost 2 hr getting dressed in their survival equipment inside the cramped confines of the Soyuz craft (Figure 7.6) before jumping out. The problem is that all the dressing is done with the hatch closed, which generates significant heat loads, to say nothing of the motion sickness induced by all the bobbing about in the waves! Finally, once all the astronauts are dressed, they jump out of the capsule and launch

Figure 7.6. Canadian astronaut, Robert Thirsk, sits inside the cramped interior of the Soyuz. Image courtesy: CSA.

signal flares. Although they are only training, due to the leak rates of the training suits, most astronauts are actually in need of being rescued at that stage!

Typical survival training week

A typical survival training week begins with two preparatory days spent in the classroom learning about the Soyuz emergency kit (наз, pronounced naz), how to build shelters, and becoming familiar with clothing and the elements of basic winter survival. After the theory lessons, astronauts are taken to a garage-like building on the Star City training area of GCTC. Here, they practice donning their Sokol spacesuits (Figure 7.7) they will wear when they leave the ISS with the Soyuz capsule at the end of the mission. Since the Soyuz has about as much space as a compact car, the donning exercise is nearly always a struggle. Once they've built up a sweat inside the Soyuz, the astronauts are taken out into the field, where they spend 2 days practicing survival skills. First, the astronauts organize their equipment, before searching for materials for firewood and to build a shelter. Next, they build a lean-to-type shelter using parachute cloth as the roof. To insulate the floor, the astronauts use wood sticks and more parachute cloth. Once the shelter is constructed, the astronauts turn their attention to food. First, they have to light a fire. The fire serves not only as a means of heating food, but also as a signal, alerting rescue forces in the event of a real emergency landing. Finally, once the astronauts have eaten, they draw straws to see who will take the first watch (many of the landing sites in Russia happen to be inhabited by wolves and bears), which is usually between 10.00 pm and 1.00 am. With the watch schedule decided, the astronauts turn in for the night, leaving the astronaut on guard to watch for hungry bears and making sure the fire doesn't go out. When morning finally rolls around, the astronauts snack on ration packs, consisting of freeze-dried chocolate, cookies, and tea flavored with sugar and lemon.

Day Two of their survival exercise is spent gathering more firewood for the fire, staying warm, and trying to get some sleep. After waking up on Day Three, the astronauts are instructed to gather their equipment and march to a rescue helicopter. It may sound like a simple exercise, but the instructors usually throw a spanner in the works by "disabling" one of the astronauts. The other astronauts have to respond by administering medical attention and build a stretcher to effect an evacuation. Finally, when they reach the helicopter, the exercise ends. For the astronauts, it's not a moment too soon!

Canada

Robotics training in Saint Hubert

After spending 4 weeks in Star City, astronauts head back over the Atlantic to Montreal, Canada, where they undergo 2 weeks of rigorous training, learning how to

Figure 7.7. Canadian astronaut, Robert Thirsk, dons the Sokol spacesuit. Image courtesy: CSA.

operate the Canadarm2 (Panel 7.4 and Figure 7.8) and Dextre (Panel 7.5 and Figure 7.9), two robot arms designed by MacDonald Dettweiler of Canada.

Panel 7.4. Canadarm2

The Canadarm2 moves end-to-end to reach many parts of the ISS in an inchworm-like movement. Its range is limited only by the number of Power Data Grapple Fixtures (PDGFs) on the station. The PDGFs are located at various points on the ISS and provide power, data, and video to the Canadarm2 through its Latching End Effectors (LEEs). The Canadarm2 is also capable of traveling the entire length of the ISS on the Mobile Base System (MBS).

The versatile arm has seven degrees of freedom (DoF) and, much like a human arm, has a shoulder (three joints), an elbow (one joint), and wrists (three joints). Thanks to its joint configuration, the Canadarm2 can rotate 540°. Force moment sensors provide a sense of touch and the arm is fitted with an automatic vision feature for capturing objects and automatic collision avoidance. It also has four color cameras, one on each side of the elbow and two on the LEEs.

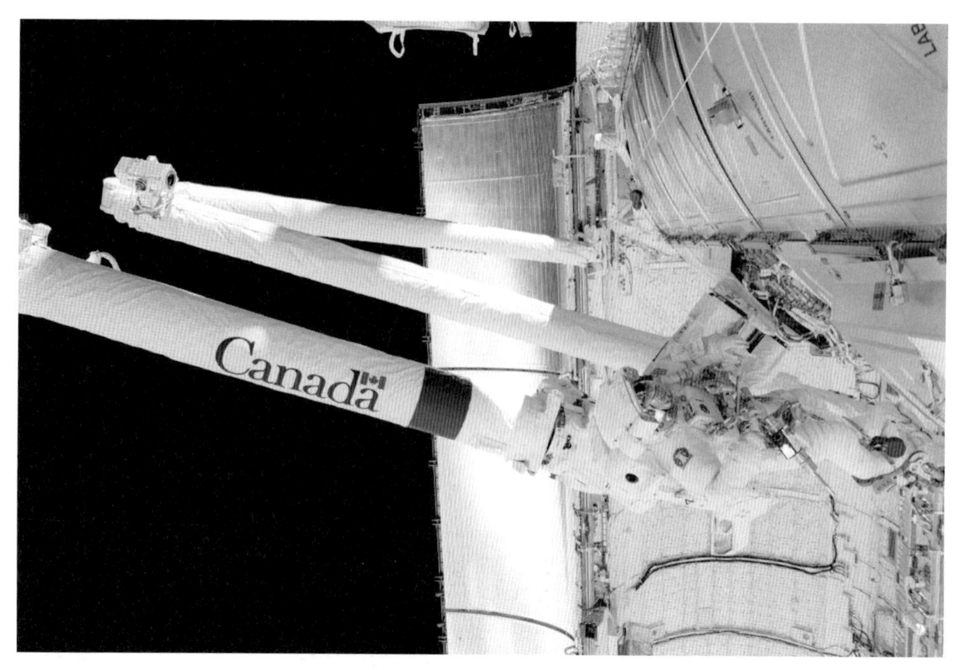

Figure 7.8. NASA astronaut, Scott E. Parazynski, mission specialist, works with cables linked to the Canadarm2. Image courtesy: NASA.

Panel 7.5. Dextre

The two-armed Special Purpose Dexterous Manipulator, known as "Dextre", is a component of the Mobile Servicing System (MSS) developed by MacDonald Dettweiler of Canada for the ISS. It complements the mobile base and Canadarm2 already installed and operating on the ISS. With its advanced stabilization and handling capabilities, Dextre can perform delicate tasks usually reserved for humans and, thanks to its versatility, it increases crew safety and reduces the amount of time astronauts must spend outside the ISS on routine maintenance, thereby freeing time for scientific activities.

Akin to a mechanic in space, Dextre can pivot at the waist, and its shoulders support two identical arms with seven offset joints, permitting great freedom of movement. It is equipped not only with lights, video equipment, and a stowage platform, but also with several robotic tools. At the end of each arm is an orbital replacement unit/tool capable of grasping a payload with a vice-like grip. For fine manipulation tasks, Dextre is capable of precisely sensing forces and torque in its grip by means of automatic compensation. To grab objects, Dextre has a retractable motorized socket wrench to turn bolts and mate or detach mechanisms, as well as a camera and lights for close-up viewing. Dextre can either be attached to the end of Canadarm2 or ride independently on the mobile base system. Thanks to its ingenious design, Dextre can accomplish tasks requiring high precision, such as removing and replacing ISS components, opening and closing covers, and deploying or retracting mechanisms.

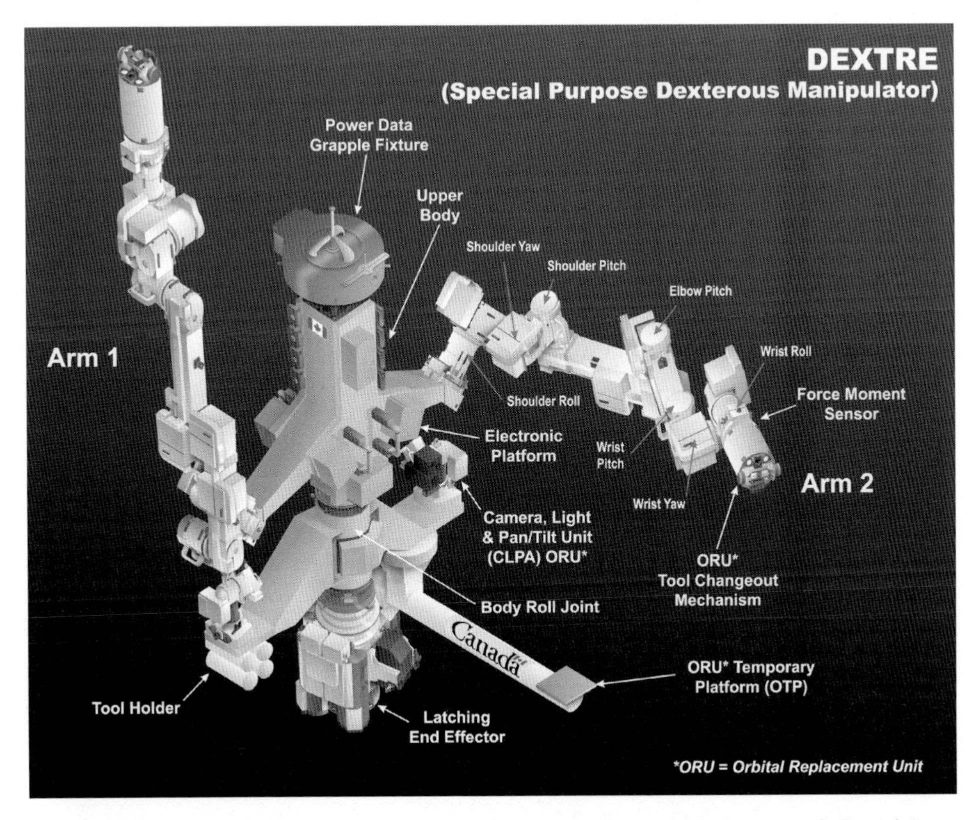

Figure 7.9. Dextre is a sophisticated dual-armed robot, which is part of Canada's contribution to the International Space Station. Image courtesy: CSA.

The training takes place at the Canadian Space Agency's (CSA's) Operations Engineering Training Facility (OETF) at the John H. Chapman Space Centre in Longueuil, a small town outside Montreal. Like so much of their training, astronauts must first go through theoretical learning, followed by practical training. In Saint-Hubert, the astronauts are taught the theoretical aspects in the Multimedia Learning Centre (MLC), before moving onto more practical training in the Mobile Servicing Systems Operations Training Simulator (MSSOTS, Figure 7.10). For those astronauts who are pilots, the pitch, roll, and yaw movements of the robot arms are intuitive and easily grasped. For astronauts without a piloting background, the instruction can sometimes be a little frustrating.

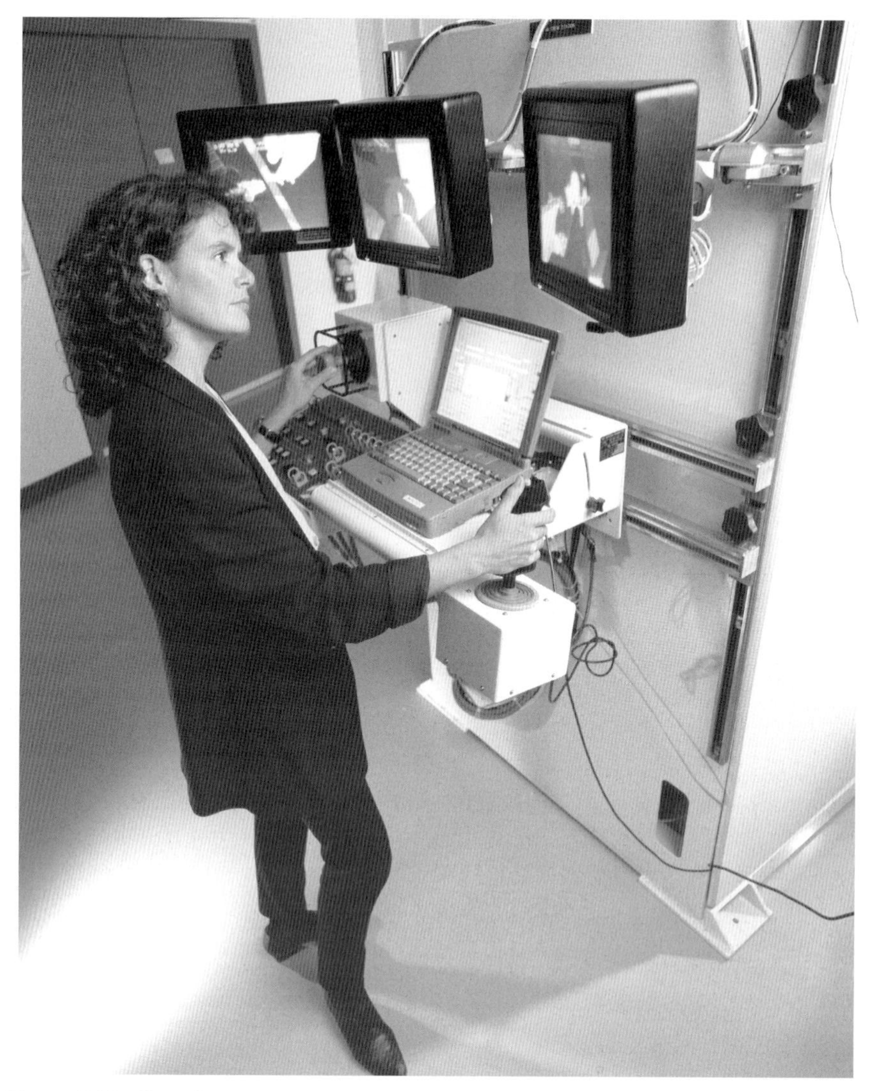

Figure 7.10. Canadian astronaut, Julie Payette, practices using the Mobile Servicing Systems Operations Training Simulator. Image courtesy: CSA (*see colour section*).

Robotic Workstation

Astronauts also train on a Robotic Workstation (RWS), located in a three-quarter-scale mockup of the US Destiny Lab. The RWS contains a Display and Control Panel, a Portable Computer System for displaying the Graphical User Interface, and two hand controllers. The hand controllers control the translations and rotations of the end of the arm. Assisting the astronauts are three monitors displaying views from cameras located on Canadarm2 and other locations outside the ISS. A special feature of the RWS is the Canadian Space Vision System (CSVS), enabling astronauts to pinpoint the exact location and movement of payloads, providing them with an accurate measure of distance and orientation, which is critical in the maintenance of the ISS.

Flying the arm

Astronauts refer to operation of the arm as "flying the arm", a reference to the arms pitch, yaw, and roll capabilities. Learning to fly the arm can be a little confusing but once astronauts imagine the robot arm as being an extension of their own arm, the whole operation becomes a little easier. The astronaut's joints, much in the same way as the robot arm, provide DoFs, and the "shoulder" and "wrist" joints move in ways comparable to the engineering equivalent. Once the astronauts have become accustomed to the arm's shoulder, elbow, and wrist, they must then grapple with how to use the hand, which, in robot terminology, is the LEE.

If you want to visualize how to operate the Canadarm2, just reach out with your right arm to pick a pen off a desk. To do this, you will extend your arm and pick up the pen by grasping it with your LEE. By grasping the pen in your hand, you will have executed a successful grapple. Now, move the pen to a position directly above your head while simultaneously rotating the pen so the point faces down. You will notice how your joints move to execute the maneuver. Next, try to move your arm over your head so the tip of the pen touches your left shoulder, while simultaneously rotating the pen so that it points behind you. Again, notice how your joints move and you have will have some idea of the complexity of the robotic arms.

One of the most difficult tasks the instructors give the astronauts is the free-flyer capture exercise. In this scenario, an object is floating freely in space with the possibility it may impact the ISS. To prevent impact, the astronaut must chase the object and secure it – a task requiring the use of complicated sensory and motor skills in a very limited amount of time (usually less than 90 sec).

Russia Part II

After 2 weeks of intensive training, astronauts are certified as Mission Robotics Operators (MROs) and they head back over the Atlantic to Star City once again for their second 4-week ISS training session. After spending 2 weeks in French-speaking Montreal, many astronauts find they have to brush up on their Russian language skills, which were pretty basic to begin with. At this stage, many astronauts have yet to master the Cyrillic alphabet, while others struggle putting basic sentences together. By now, for most astronauts, the realization has dawned on them that the challenges of the Russian language are as formidable as anything they will encounter during mission training.

Soyuz training

While still coming to grips with the intricacies of speaking Russian, training continues in the operation of Soyuz systems. This means the astronauts must study literally reams of technical data (all in Russian!) and attend a seemingly endless series of theoretical classes (again, in Russian!). The first systems astronauts study is the Soyuz's docking and attachment system, the thermal control system, and the electrical power and propulsion system. The instruction covers all aspects of the active and passive elements of the thermal control system, the air conditioning and ventilation systems, and the fire suppression system. In common with all training, each system is taught in a theoretical lecture (*lek-see-ya*) followed by a practical (*prak-ti-ka*) session. The theoretical session consists of an instructor talking while the astronauts take notes, while the practical sessions, which the astronauts enjoy more, are hands-on. In the practical sessions, the astronauts have the opportunity to touch hardware, flip switches, read gauges, turn dials, and even enter commands into computers. Once the lectures are over, the astronauts typically go straight home to study all the documentation accumulated that day and on previous days. The endless studying is necessary because, eventually, the astronauts are tested on what they have been taught. Before the exam, just like at university, there is a review session in which astronauts can ask any question to clarify concepts about the systems. Occasionally, the system designers from Energia attend the review sessions, which is helpful, since these engineers are the same ones who ask the questions in the exam.

In the exams, which usually last an hour, the astronauts are asked various questions to demonstrate their knowledge of the system and occasionally are asked to perform various functions within the simulator. On completion of the exam, the engineers rate the astronauts on a scale of 1 (fail) to 5 (excellent).

Spacesuit fitting

After acing the systems exams (most of them do), astronauts are fitted for their seats on the Soyuz. The suit fitting begins with a 60-km drive to Zvezda Space Facility (ZSF), where the spacesuits are manufactured. First, astronauts are given a package containing three sets of underwear, three pairs of socks, and three sets of knee-length long johns. Then, they're sent to the bathroom and told to change into the briefs and socks with the long johns over the briefs. As soon as they've changed, a team of five engineers surrounds them and starts taking measurements. After the measurements are completed, the astronauts are placed in a plastic mockup of a Soyuz seat. Wearing their less-than-chic underwear, the astronauts are then placed in another mockup seat, which looks surprisingly similar to a small bathtub. Once in the supine position, the engineers start pouring fast-drying plaster around the astronaut, with the intention of forming a mold around their body – a process that eventually leads to building a custom seat liner (Figure 7.11). Not only does the custom liner fit the astronaut's body contours when wearing his/her Sokol spacesuit, but it also provides a snug fit during the dynamic launch loads and, upon return to Earth, the potentially

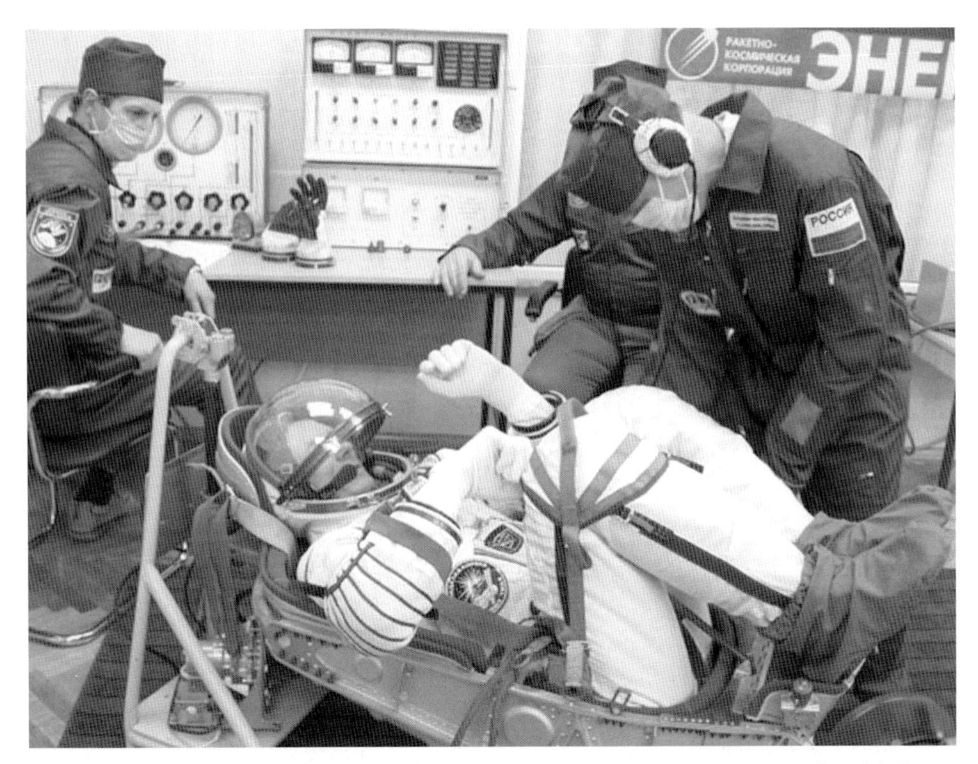

Figure 7.11. Engineers pour plaster into the Soyuz seat liner to ensure a comfortable fit. Image courtesy: NASA.

violent ground landing. After the engineers pour plaster around the upper body, they wait for it to dry before asking the astronaut to get out of the tub, after which they carve and smooth the plaster form to relieve any potential pressure points. The process is then repeated a number of times as the engineers smooth the final shape until finally they are left with a full-size final mold. After the seat liner fitting, and changing into new underwear (!), the astronauts have an opportunity to don the Sokol suit, an experience that makes the dream of flying into space appear just a little closer.

After their second phase of training in Star City, astronauts once again head back over the Atlantic to Johnson Space Center (JSC), where they are introduced to the challenges of conducting extravehicular activity (EVA) in the world's largest swimming pool.

United States Part I

Neutral buoyancy training

Measuring 62 m in length, 31 m in width, and 12 m in depth, the Neutral Buoyancy Laboratory (NBL) is filled with 22.7 million liters of water. At the bottom of the

Figure 7.12. An astronaut entering NASA's Neutral Buoyancy Laboratory. Image courtesy: NASA.

NBL (Figure 7.12) sits a full-size mockup of the ISS, which is why the NBL needs to be so big! Because the facility is used to train ascans to scuba-dive, the NBL is familiar to astronauts, but the real purpose of the oversized pool is to train crewmembers in the art of EVA.

Astronauts learning how to perform EVAs use a bulky training suit. Because the suit is so large, astronauts spend a lot of time simply practicing moving around while wearing it and learning how to use tools with bulky gloves on their hands. Due to the cumbersome suits and the complex tasks astronauts must perform during EVAs, the training often occupies a significant proportion of mission preparation. In fact, astronauts usually spend 7 hr training underwater for every hour they spend spacewalking during a mission.

Being underwater in the NBL is similar to, but not quite the same as, being in space because the astronaut is not truly weightless as in space. Instead, they are what scuba-divers refer to as neutrally buoyant. Being neutrally buoyant means an object doesn't want to float to the surface or sink to the bottom. Scuba-divers try to be neutrally buoyant so they don't sink or float upwards when they're underwater. In the NBL, to help astronauts achieve neutral buoyancy, divers attach weights to the relatively buoyant spacesuit. With the right amount of ballast, astronauts feel much like they will in space.

Before astronauts don their training spacesuits, they must first practice their EVAs while wearing standard scuba gear. Once they're comfortable with their assigned tasks, they start practicing them while wearing the spacesuit. In the NBL, scuba-divers (Figure 7.13) help the astronauts move around until they get used to moving in their spacesuit. The divers are also there to protect the astronaut in case they have a problem with the suit. Although the NBL training is one of the most exhausting types of training, it is also one of the most enjoyable because astronauts get to experience the wonderful sensation of being weightless, albeit simulated!

Once the astronauts learn how to stay in one place (if they push too hard in space, they will simply float away!), they are taught how to use all the tools they will need during their spacewalk. The effects of weightlessness, combined with the cumber-

Figure 7.13. Divers assist astronauts during their simulated EVAs. Image courtesy: NASA (*see colour section*).

some EVA suits, mean the task of manipulating tools can be extra challenging, as STS-126 astronaut Heide Stefanyshyn-Piper discovered (Panel 7.6) in November, 2008.

Panel 7.6. Tool bag overboard

During Space Shuttle *Endeavour*'s mission STS-126 in November, 2008, NASA astronaut Heide Stefanyshyn-Piper was cleaning up the mess from a leaky grease gun when the small bag came loose and drifted beyond her reach with its load of vital (and expensive!) tools. Stefanyshyn-Piper was conducting an EVA outside the ISS with fellow astronaut, Steve Bowen, to clean grit out of a damaged solar gear and add lubrication to restore its health. The gear is used to turn the station's starboard solar arrays so they always face the Sun.

It was while preparing for the EVA that Stefanyshyn-Piper, a veteran spacewalker, opened the tool bag and found it full of sticky grease. The grease, Stefanyshyn-Piper described, filled the bag with thick, goopy pieces that stuck not only to tools, but also to her spacesuit. While she was using wipes to clean up the mess, the 13-kg bag (containing $100,000 of tools) popped loose and started floating away! For a moment, Stefanyshyn-Piper thought she might be able to jump and grab it, but quickly realized that would have made everything worse, since there would have been two floating objects, one of which would have been her! To be fair to Stefanyshyn-Piper, Mission Control believed the bag might not have been secured properly from the start. Also, Bowen took some of the heat, since it was he who prepared and cleared the bag for use before the spacewalk. Nevertheless, there was some good-natured ribbing among *Endeavour*'s crew following the event.

Regardless of what EVA tasks they are required to perform on orbit, the astronauts must practice every task dozens of times in the NBL, under the careful supervision of the divers until they can do it correctly *every* time. Since every astronaut will be a member of the ISS crew, they must also be trained to perform numerous tasks, which is another reason NBL training is often so arduous. In most cases, the tasks taught are not planned to be conducted in orbit, but are taught in case something breaks outside the station. Because of the myriad systems that could conceivably fail, it is necessary to train astronauts to be a jack of all trades, which means an EVA astronaut is often working as an EVA repairman.

Challenges of EVA

The most difficult part of EVA training, astronauts will tell you, is the length of time it takes. Including the preparation period, a typical EVA training session can easily stretch to 9 or 10 hr (Table 7.3). Another challenge is simply trying to keep track of all the tools (almost 100!) an astronaut may use in a session. Then, there are the complexities associated with trying to complete what would be an easy task if it weren't for the bulky gloves and compounding problems of lack of gravity. For example, the simple act of unfastening a bolt might sound straightforward, but each astronaut will perform the task differently, depending on their strength, skill, and fatigue level. Because no EVA procedure is simple, EVA training staff spend hours discussing a session with astronauts before a training session and hours debriefing the crew afterwards. Because of the complex nature of EVA, it is a favorite topic of discussion among astronauts, with most agreeing that "slower is faster" is the most effective tactic.

Table 7.3. Daily dive operations in the Neutral Buoyancy Laboratory.

Time	*Event*
05.30–06.30	Critical system startup
06.30–07.30	Pre-test checklist (test team: 50 personnel)
	Physical training for staff (lap swim)
	Suit and tool delivery to tank deck
07.30–08.00	Astronaut pre-brief by test conductor
	Daily physicals
	Final test setup preparation (2nd dive team, Rotation 1)
08.00–08.15	Test team pre-dive briefing
08.15	Test team call to station
08.15–08.40	Suit donning
08.40–09.00	Subject immersion and initial weigh-out
09.00	Transition of control of test from test director to test conductor
09.00–15.00	Subject test procedures
09.00–11.00	1st dive team, Rotation 1 (four divers/subject: two safety, one utility, one camera)
11.00–13.00	2nd dive team, Rotation 2
13.00–15.00	1st dive team, Rotation 2
15.00–15.15	Subject recovery and suit doffing (10-min post-dive observation period)
15.15–15.30	Locker room
	Suit and tool recovery and servicing
15.30–17.00	Debrief
15.30–18.30	Reconfiguration shift, first dive (3 hr)
20.00–22.00	Reconfiguration shift, second dive (2 hr)
22.00–23.00	Post-dive observation period (OSHA requirement)

NASA suit fitting

In between their NBL training sessions, astronauts must also be fitted for a NASA suit. In contrast to the "one size fits all" limitations of the Russian Sokol suit, NASA customizes its suits. The suit fitting is a relatively straightforward procedure, requiring astronauts to strip down to their underwear, don some spandex, and be shot at with laser beams for 3 hr! NASA also utilizes this laser technology as a means to gather data on the astronaut's physical measurements. By creating a database, scientists and engineers can analyze the data and determine the different body shapes, heights, arm lengths, and hand sizes of those selected to fly in space. This, in turn, will not only enable better design in the development of new spacesuits, but will also aid in the development of spacesuits, providing a more uniform fit and increased comfort and functionality. The intent of the designers is to forgo the traditionally bulky, stiff, heavy, and task-specific designs in order to create a suit, such as the BioSuit (see Chapter 6), that is more flexible, lighter-weight, and easier to maneuver and work in.

Survival training in the US

Following suit fitting and NBL training, astronauts spend 2 weeks survival training to refresh skills learned while they were ascans. The training begins with an orientation of the training facilities and a meeting with the instructors. This is followed by equipment issue and a drive to the training area. If they are being trained in winter survival, the astronauts are equipped with backpacks, skis, and sleds, and

Figure 7.14. The Absaroka Mountains serve as an ideal location for astronauts to practice survival training. Image courtesy: Wikimedia.

dropped off in the wilderness, such as the Absaroka Mountains (Figure 7.14), a sub-range of the Rockies, with dozens of mountains higher than 3,000 m. Here, the astronauts hike up some of the smaller mountains and learn how to build snow shelters, understand avalanche terrain, rescue other crewmembers trapped in avalanches, and navigate using maps, compasses, and Global Positioning Satellites (GPS).

In addition to dodging avalanches, the astronauts learn how to cook outdoors and become familiar with the myriad skills required to stay alive in the winter. Once the instructors are satisfied the astronauts have mastered at least the basic skills, they leave them to fend for themselves. For the next few days, the astronauts move to and from their base camp with the objective of locating caches of food and fuel. With temperatures of no more than –10°C and wind chills making it feel twice as cold, the tasks are often accomplished only with considerable discomfort. On the coldest days, the astronauts build snow shelters, forcing them to live like gophers in holes burrowed into the snow! Finally, after having survived more than a week in the open, the astronauts trek the 10 km back to the road to be collected by their instructors.

Some may wonder why astronauts spend so much time on survival training. Surely, with all the technology available on American and Russian spacecraft, astronauts should be able to make a pinpoint landing anywhere on Earth? Unfortunately, it's not that simple. More than once, the trajectory of a returning Soyuz has gone awry and cosmonauts have found themselves many kilometers from the planned landing site. In fact, given the remoteness of much of Russia and the local wildlife, Russian Soyuz capsules routinely carry guns so downed cosmonauts can defend themselves in the event of encountering hungry bears or wolves. In case you think this is a little far-fetched, the practice of issuing guns (Figure 7.15) on spacecraft started in 1965 after a Soyuz landed in a remote area of Russian territory. Apparently, the crew encountered hungry wolves, although, reportedly, the rescue helicopter team stated the wolves were about 3 km away from the landing site and nowhere near the capsule. In contrast, NASA has never carried firearms into space, although machetes have been provided in case a capsule were to land in a jungle area.

Russia Part III

Orlan spacesuit

On their return to Star City, the astronauts look forward to training on the Orlan spacesuit. Unlike the Sokol suit, which is designed to be worn only during launch and entry, the Orlan-M is Russia's spacewalking suit. Whereas NASA astronauts perform EVAs, Russian cosmonauts perform spacewalks or выход в открытый космос (literally, to walk in open space). Also, whereas astronauts use an extra-vehicular mobility unit, or EMU (Figure 7.16), to perform their EVAs, the cosmonauts use the Orlan-M (Figure 7.17).

Figure 7.15. Canadian astronaut, Robert Thirsk, practices firing a weapon in preparation for his upcoming mission. Image courtesy: CSA (*see colour section*).

Figure 7.16. NASA's Extravehicular Mobility Unit. Image courtesy: NASA.

Figure 3.4

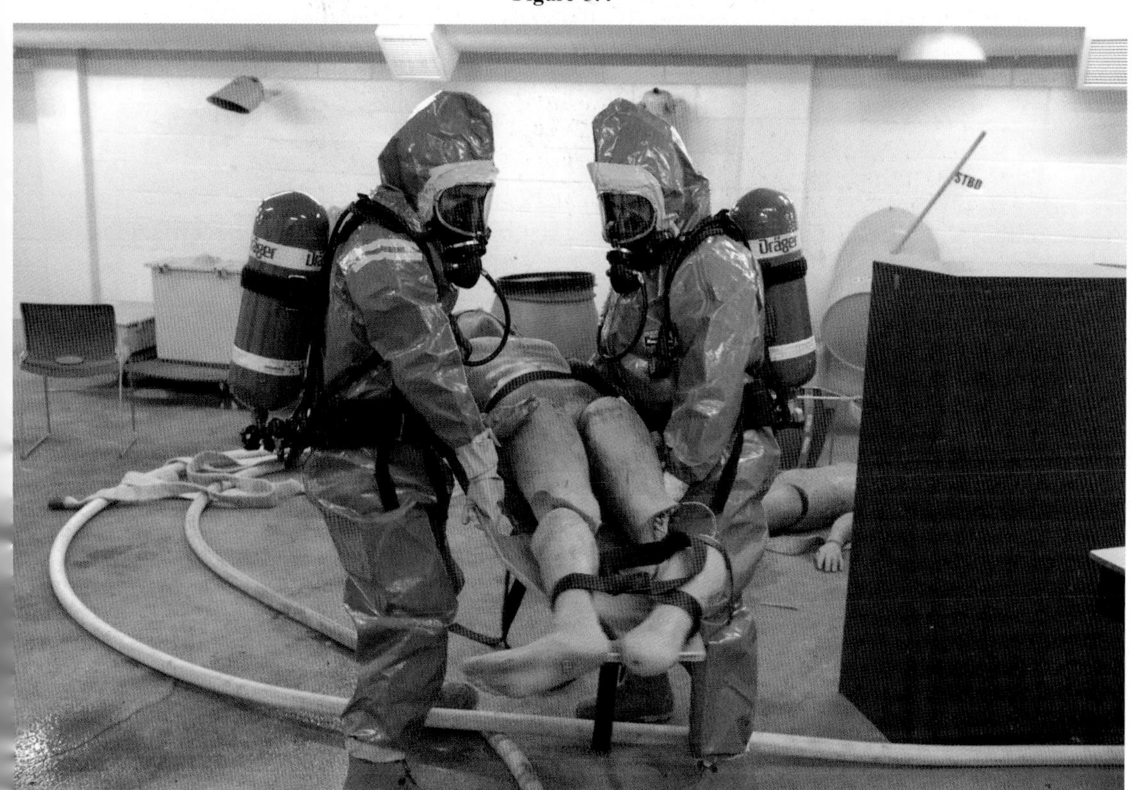

Figure 3.6

Figure 3.9

Figure 3.11

Figure 4.8

Figure 4.14

Figure 4.18

Figure 4.20

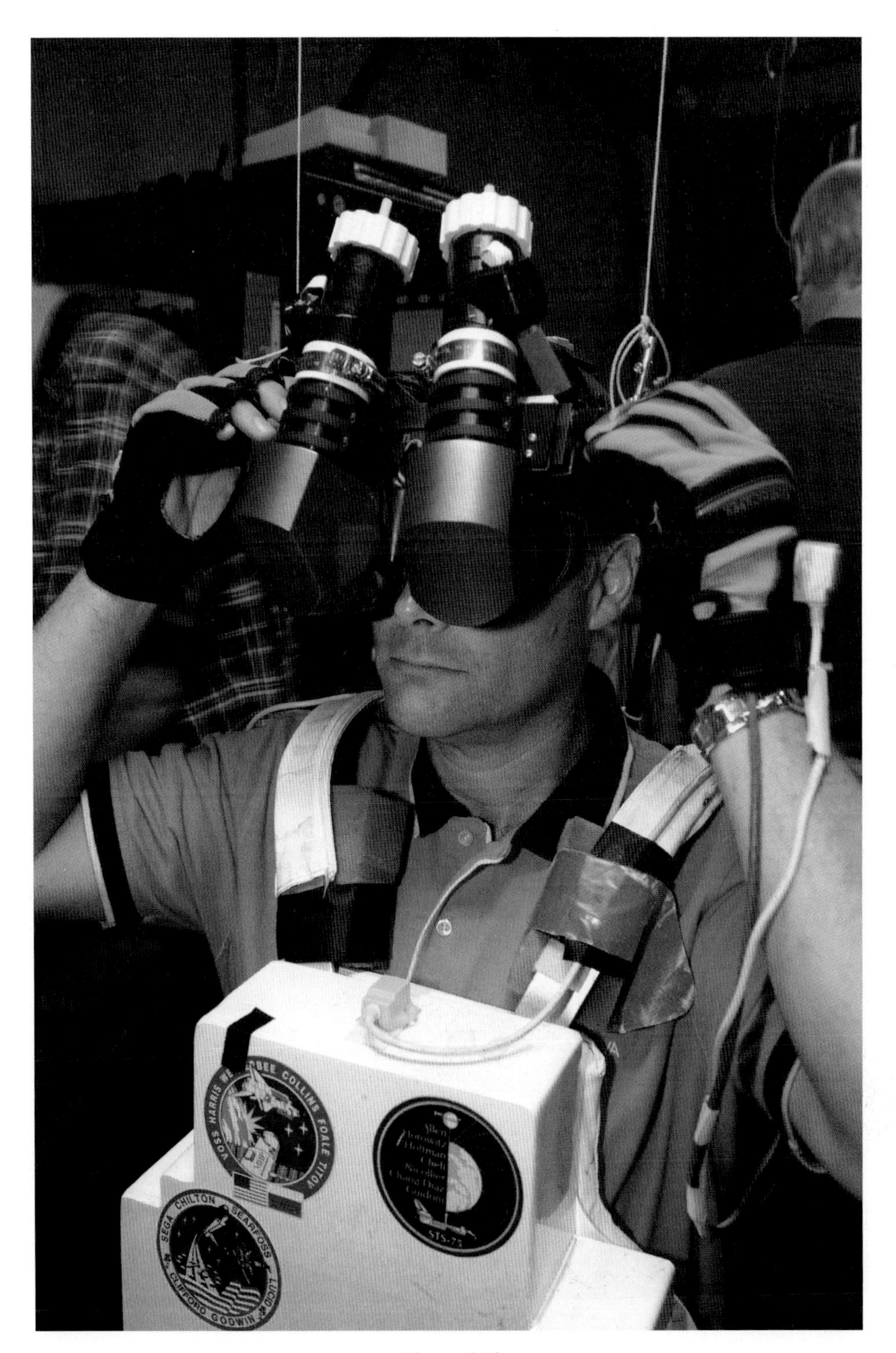
VOSS HARRIS
COLLINS FOALE TITOV
SEGA CHILTON SEARFOSS
CLIFFORD GODWIN LUCID

Figure 6.13

Figure 6.14

Figure 7.5

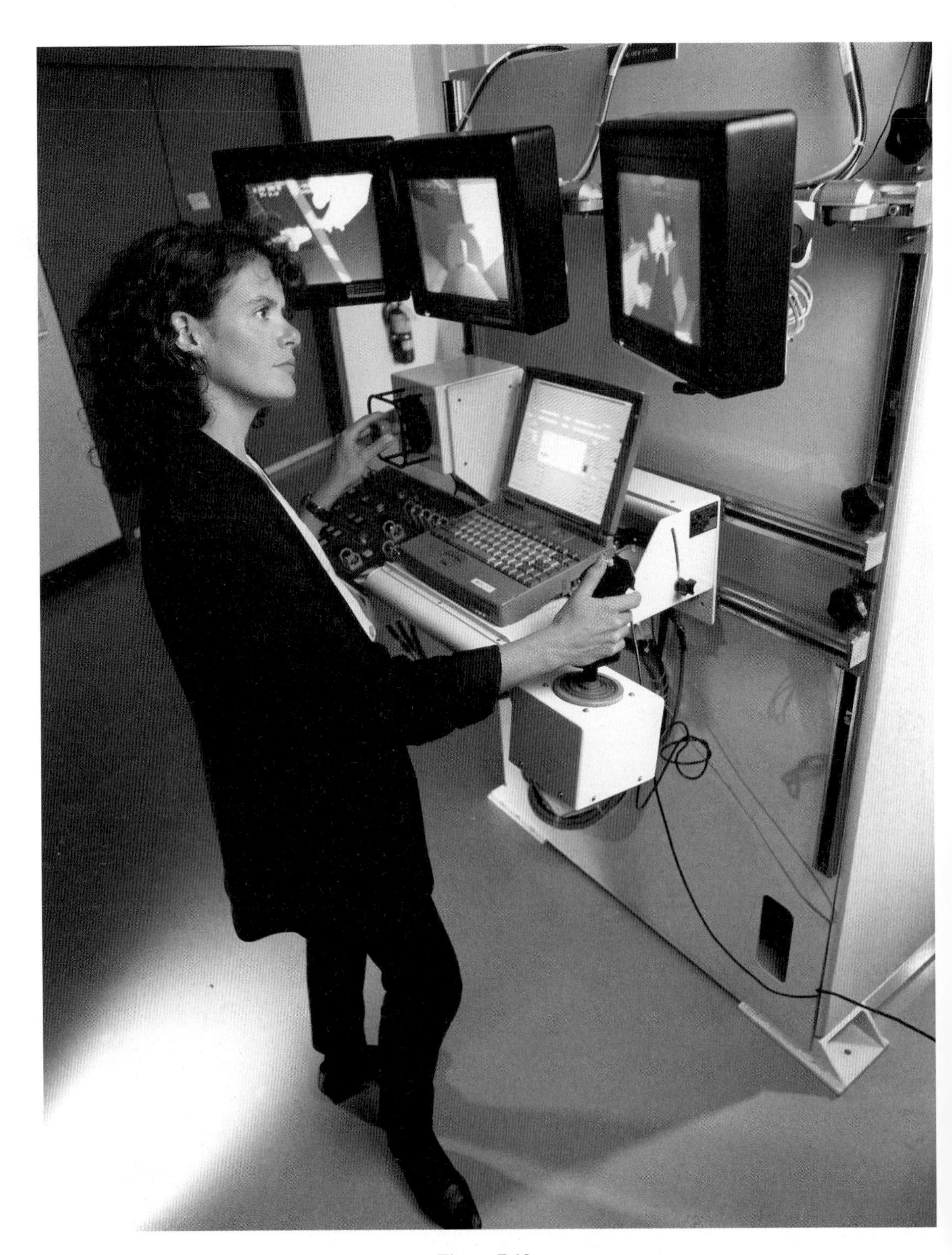

Figure 7.10

Figure 7.13

Figure 7.15

Figure 7.18

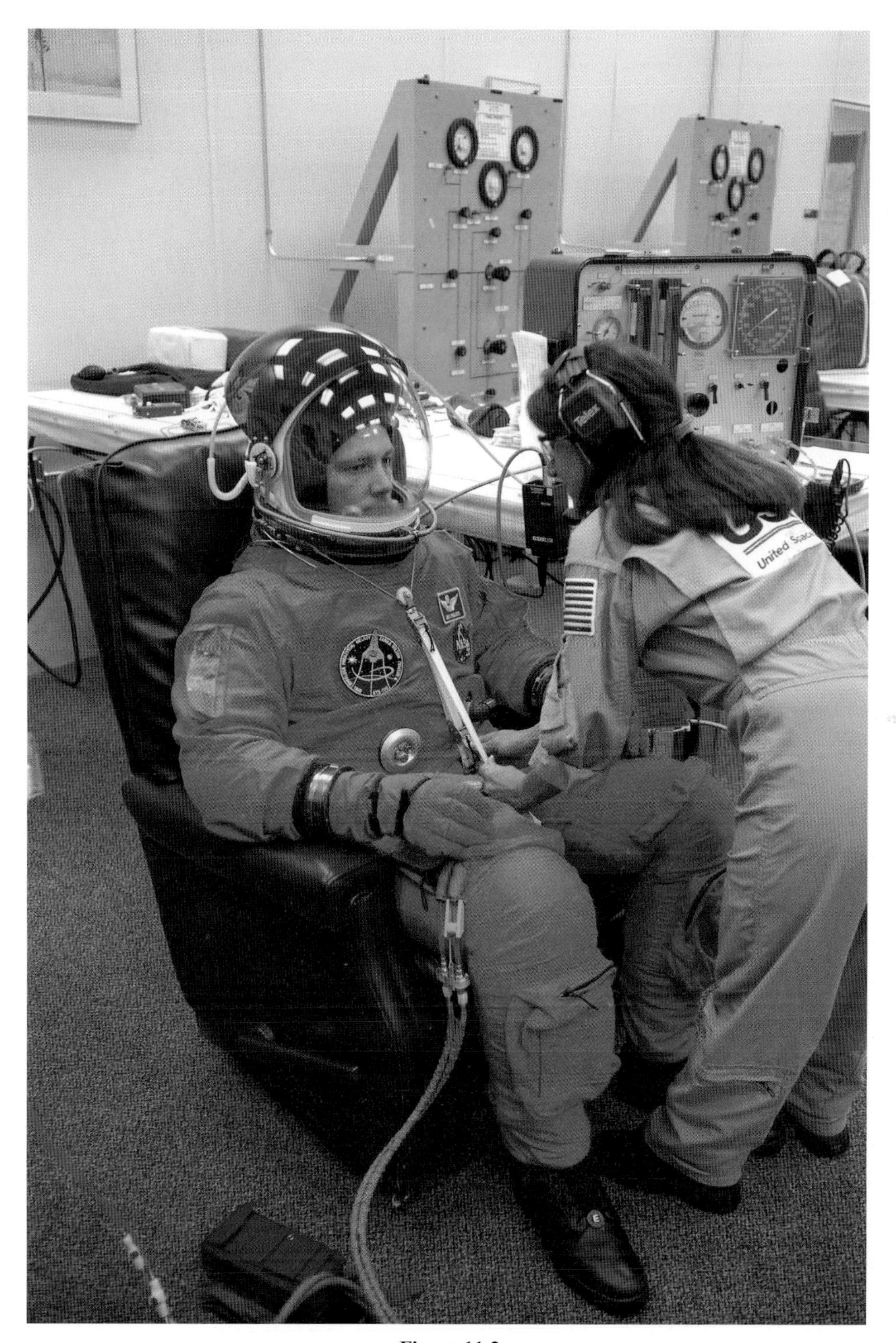

Figure 11.2

Figure 11.5

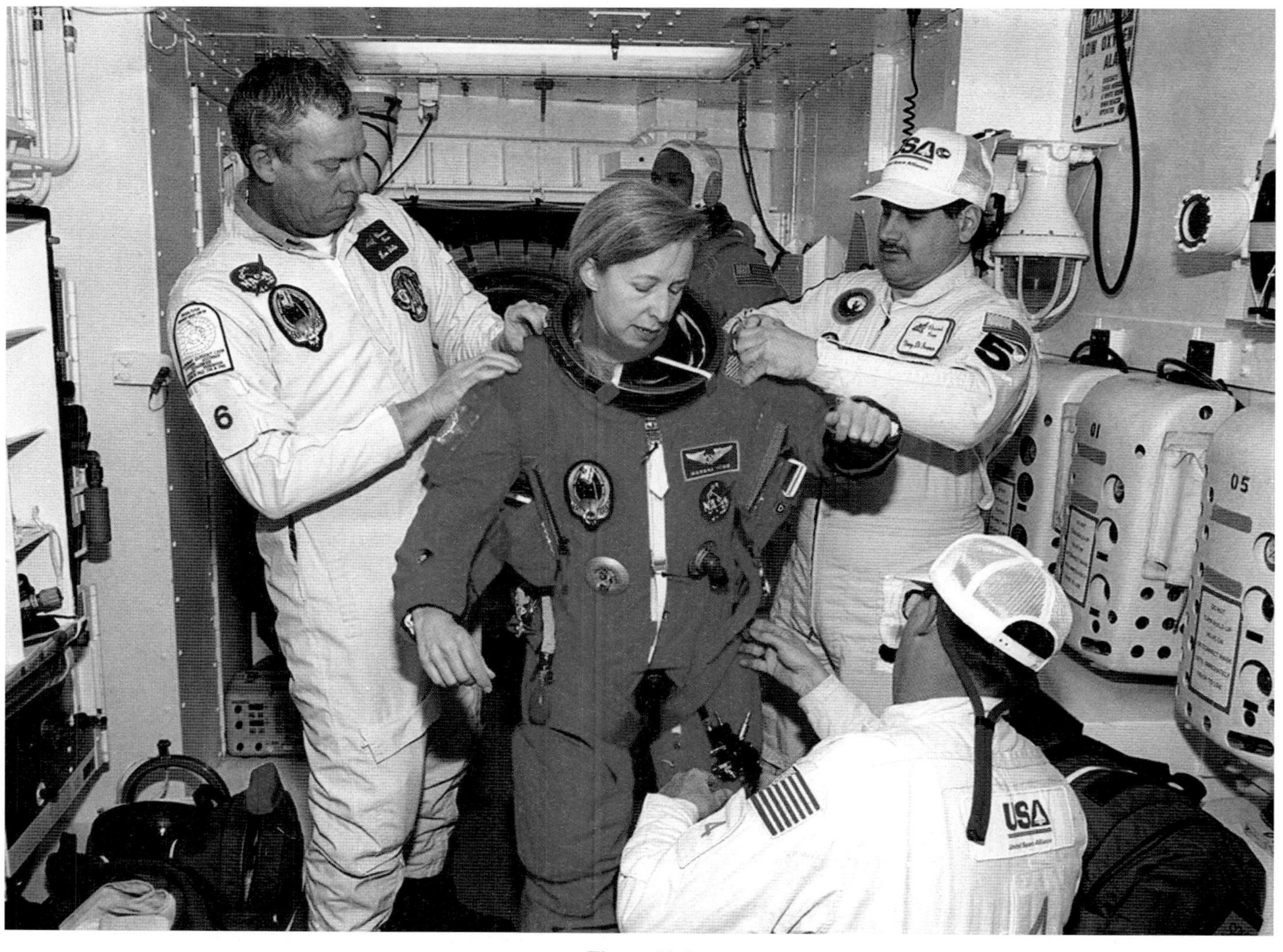

Figure 11.6

Figure 11.7

Figure 7.17. The Russian Orlan-M spacesuit. Image courtesy: NASA.

While the Russians use different terminology and spacesuits, their approach to training astronauts and cosmonauts in the skills required to perform EVAs is much the same as the Americans. In Houston, EVA training is conducted while wearing a training EMU in the NBL, whereas the Russians have their own pool, called the Hydro Lab. To make training appear as realistic as possible, the Russians suspend their cosmonauts, thereby providing them with an approximate simulation of zero gravity. Helping with the realism is the mockup of the Russian ISS segment airlock, which sports a hatch enabling astronauts to practice some basic EVA techniques while wearing the Orlan. These techniques include depressurization/re-pressurization procedures, opening/closing the airlock hatch, use of tethers, and, most important, executing steps used in the event of equipment malfunctions.

Unlike NASA's EMU suit, which comes in various sizes, the Orlan is a "one size fits all" suit. For some astronauts, it makes for a tight fit, although the suit can be sized by lengthening the arms and legs. Unfortunately, there are only two sizes of gloves, which most astronauts refer to as size small and very small!

Food tasting

Yes, food tasting is a training session! Unlike the NASA version, in which astronauts are presented with actual meals (appetizers followed by main courses, drinks, and desserts), the Russians present their food in different groupings, one for each day of tasting, with drinks interspersed between each session. In addition to giving astronauts an appreciation for Russian cuisine, the food tasting allows crewmembers to express food preferences for their trip into space. The process is quite simple. After each food sample, the astronauts complete a documentation sheet and record a rating for the food. A "nine" rating indicates the astronaut likes the item, whereas a "one" rating means the astronauts probably won't be including the item during his/her mission! After rating the food item, the astronaut is asked to estimate how many times they might want to eat it during a food cycle and list any comments they have. Not surprisingly, Russian gastronomy is rather different from the fare American astronauts are used to. During a tasting session in Star City, they will typically try everything from mashed potatoes with meat to chicken with prunes! The food tasting session lasts 3 days, after which the astronaut's taste buds are little tender. Later during their training, the astronauts have the opportunity to eat an entire Russian menu, spread out over several training days, based upon their food ratings. Onboard the ISS, they will receive their food in a 10-day rotation, but they're not restricted to eating just potatoes and meat because they have the option of eating either two American meals and one Russian, or two Russian and one American.

Living away from home

Inevitably, with so much time spent away from home, the stresses of dealing with separation become more and more challenging. Unlike a military deployment, where soldiers come back home after having spent a year away, for astronauts, the end of training signifies the beginning of a 6-month mission and even more time away! At this stage of training, having spent several months away from family and friends, many astronauts recognize that the most difficult obstacle to overcome is not learning Russian or wearing diapers, but handling the personal trials that come with being separated for months on end. For all astronauts, the time away from home is especially difficult, unless you happen to be a Canadian, whose employment contract is a little more family-friendly than their American counterparts. A Canadian astronaut deployed abroad is entitled to a free visit from his/her spouse every 21 days, a privilege that certainly helps the marital situation when one considers many astronauts spend less than 10 weeks at home in any given year!

Forel suit

After learning how to shoot wolves and bears, astronauts are indoctrinated into the use of the colorful Forel suit (Figure 7.18). The Forel (Russian for "trout" incidentally) suit is used in the event the Soyuz lands in freezing water. In training for water landings, astronauts are taught to jump from the Soyuz in the suit, with water wings and Forel suit in tow. The one-piece rubber suit with gloves and a hood seals the astronaut's body and keeps water out. To survive other types of weather and terrain, astronauts simply add or remove some of the multiple layers to keep warm

Figure 7.18. Canadian astronaut, Robert Thirsk, models the colorful Forel suit. Image courtesy: CSA (*see colour section*).

or cool, depending on the conditions they encounter when they exit the capsule. Fortunately for the astronauts, Russian search and rescue forces guarantee a pickup within 48 hr!

Pulling Gs

During the launch, ascent, orbital, de-orbit, and re-entry phases, astronauts encounter different acceleration stresses. For example, during a nominal re-entry of the Soyuz spacecraft, astronauts will experience between 4 and 5 G. Although the spacecraft has been designed to reduce these stresses, in the event of a contingency, especially during re-entry, these forces may be very large (Panel 7.7), which is why G-training in the centrifuge is included in the astronaut's training schedule.

G theory

The sustained acceleration (+Gz) an astronaut experiences during launch and re-entry is acceleration that lasts for more than 1 sec and is a force that can make it

Panel 7.7. Ballistic re-entry

In October, 2007, a Soyuz TMA-10 landed in Kazakhstan, bringing outgoing ISS commander Fyodor Yurchikhin, flight engineer Oleg Kotov, and Malaysia's first man in space, Sheikh Muszaphar Shukor, back to Earth after a steeper-than-usual descent. Plunging back to Earth from west to east over central Kazakhstan, the flight plan called for a landing near the town of Arkalyk. But, for some reason, the Soyuz flew a steeper-than-planned trajectory and landed short of the intended touchdown point, subjecting the crew to G-forces in excess of 8 G. It was the first "ballistic" re-entry since a Soyuz returned on May 3rd, 2003, with the ISS's sixth crew.

The following year, in April, a similar event occurred when a Soyuz TMA-11 also landed following a ballistic re-entry. An examination of the craft's flight path indicated a catastrophe had been narrowly avoided and total disaster had been prevented not by the crew, but by the robust construction of the spacecraft. Some reports indicated the Soyuz hit 11 G for a few seconds, and one of the astronauts, Peggy Whitson, recalled seeing 8.4 G on the spacecraft's G-meter.

almost impossible for astronauts to breathe. High rates of sustained acceleration can also result in blood pooling to such a degree that it may cause astronauts to convulse and eventually black out. Given the serious consequences of these events, it is important astronauts become familiar with the effects so they are able to deal with inflight events, such as gray-out, blackout, or even unconsciousness.

During their trip to and from orbit, astronauts experience five distinct phases of accelerative stress, each differing in magnitude and duration:

1. *Launch.* Typically between 3.5 and 4.5 G.
2. *Orbital.* The centrifugal force of the spacecraft balances the gravitational force, thus producing a microgravity environment of zero-gravity!
3. *Re-entry.* Acceleration stresses begin at an altitude of 75,000 m due to the sudden drag and deceleration during re-entry into the denser atmosphere. The magnitude of G-forces experienced depends on the spacecraft's angle of entry into the atmosphere. High re-entry angles ($>10^\circ$) produce very large forces (>25 G) whereas shallow angles of less than 1° usually result in forces of less than 5 G. In a ballistic re-entry of the Soyuz, G-forces may exceed 8 G!
4. *Landing.* Thanks to the Soyuz's use of big parachutes and the fact it usually lands on soft terrain, astronauts normally experience landing forces no greater than 5 G.
5. *Emergency egress.* The forces experienced during an emergency egress will be different in different phases of flight, but astronauts can expect high magnitude accelerations exceeding 15 G, sustained for 1 or 2 sec.

The physiological system most sensitive to G is the cardiovascular system (CVS). To monitor CVS changes, astronauts are instrumented with ECG and heart rate monitoring equipment so they can see for themselves how they react to increasing G. Generally, they can expect their heart rate to increase in correlation with increased G due to the acceleration force effect and the general psychophysiological stress syndrome associated with exposure to acceleration. In fact, most people experience an initial cardiovascular response, even before the start of the run, due to the anticipation of the event!

Many of the central nervous system (CNS) effects of G are a direct consequence of the CVS effects. This is because a regular blood supply is required for the CNS to function, so the ability of the astronaut's body to tolerate acceleration is related directly to adequate blood flowing to their brain. Because of this relationship, symptoms that relate to insufficient blood flow to the brain are used to determine tolerance to G. The normal index of defining G-level tolerance is to use loss of vision (LOV) in an upright-seated position at a specific level of G exposure. The visual symptoms astronauts experience during their centrifuge runs are caused by a reduction of blood flow to the retina of the eye, which, in turn, is caused by a reduction in driving pressure and higher intraocular pressure. Table 7.4 summarizes the sensory symptoms astronauts may experience during their runs.

Table 7.4. Categorization of light-loss criteria.

Symptom	*Description*	*Onset of symptoms*	*Criteria*
Gray-out	Partial LOV. Often occurs as first physiological effect of sustained G loads. Low blood oxygen levels cause peripheral vision to fade. Objects in center of FOV can be seen but seem surrounded by gray haze	3.5 G	100% Peripheral Light Loss (PLL) combined with 50% Central Light Loss (CLL)
Blackout	Gray haze envelops entire FOV and almost immediately becomes black. Astronaut will be conscious but unable to see	Above 5 G	100% CLL, but sufficient blood reaches brain to permit consciousness and hearing
Gravity Induced Loss of Consciousness (G-LOC)	Follows quickly after blackout with sustained G load. Astronaut will be unconscious but will regain consciousness when G load is released	Above 5 G	Normally occurs following increase of acceleration after blackout

LOV, Loss of Vision; FOV, Field of Vision.

Figure 7.19. Star City's centrifuge. Image courtesy: Russian Space Agency.

Astronauts are able to measure light loss by watching a light bar placed in front of them at eye level. The bar has a green light at each end and a red light in the center. When the astronaut looks directly at the light bar without moving his/her eyes or head and cannot see the green lights, but can see the red light, 100% PLL has occurred.

Centrifuge training

Training for G means being trained in Star City's mother of all centrifuges (Figure 7.19). In common with so much of the training, centrifuge indoctrination begins with a review of major theoretical elements of G tolerance.

During their introduction to the facility, astronauts are shown the interior of the gondola and the instructor points out the adjustable rudder pedals provided for foot support, and the shoulder and lap harnesses that secure the passengers. The astronauts also have the opportunity to don the facemask they wear during the run to monitor their breathing and to enable two-way communication with the console operator. When they sit in the chair, they may notice a small video camera, which records the astronaut during the run.

Shortly after breakfast on test day, the astronauts observe a dry run from the console room. The console operator then reviews the G onset loads and the operation of the communication system before assigning each astronaut to a centrifuge rotation. Instructors then explain the safety procedures and the roles of the flight surgeon and centrifuge operator before a final review of the training. Following the review, support personnel supervise the astronaut's ingress into the gondola, where they are connected to biomedical instrumentation including a 12-lead ECG, blood pressure cuffs, and respiratory monitoring equipment. Then they spin the "fuge", using profiles similar to the ones described in Table 7.5.

Table 7.5. Run schedule for determination of G-sensitivity.

Run number	*Type of run*	*Rate of onset*	*Peak G*		*Rate of offset*
		G/sec	Magnitude	Duration at G (sec)	G/sec
1	Warm-up	0.1	6.0	5	0.2
2	GOR[1]	0.1	5.0	5	1.0
3	ROR[2] 1	1.0	3.0	20	1.0
4	ROR 2	1.0	4.0	15	1.0
5	ROR 3	1.0	5.0	15	1.0
6	GOR	0.1	8.0	5	1.0

[1] Gradual onset run: 0.1 G/sec. [2] Rapid onset run: 2.5 G/sec.

While astronauts tolerate the gradual onset rates (GORs) without too much effort, the rapid onset rates (RORs) are a different matter. For the final test, the centrifuge operator cranks it up to 8 G. While most astronauts describe the experience akin to an elephant sitting on their chest, the Russian instructors prefer to use the analogy of hugging a big grandmother!

United States Part II

Gs in the T-38

Back in the US, astronauts have the opportunity to experience a more pleasant way of experiencing G in NASA's T-38 training jet, which astronauts use for spaceflight readiness training. Since the T-38 can easily fly beyond Mach 1, pulling Gs by performing barrel rolls, loop-the-loops, and aileron rolls is fairly easy.

Science preparation

About a year into the training, more mission details are finalized, such as the intravehicular activity (IVA) schedules, the actual EVAs, and crew responsibilities.

In addition to refining schedules, astronauts spend more time learning how to perform science. Since each crewmember is assigned to perform science experiments, it is important they know how to perform them accurately. To help the astronauts, teams of experts spend hundreds of hours to ensure each crewmember has the knowledge and skills needed to perform the assigned experiments.

Science onboard the ISS

Previous ISS crews have grown human cells to study how cancers grow and worked with antibiotics to find a way to produce them faster on Earth. They've grown plants to develop drought-resistant crops and crystals to improve gasoline production. They've also studied the human body in microgravity, gathering information on everything from how the lungs perform to the formation of kidney stones and the performance of liver cells. Other experiments take advantage of the microgravity environment on the ISS to study physical processes. By eliminating gravity, researchers can better understand some of the smaller forces that occur in such processes as semiconductor production. While some ISS experiments only require crewmembers to start and stop them (e.g. crystal growth studies), other experiments require the crewmembers to be operators (Panel 7.8). For example, human life sciences experiments are unique, since they require crewmembers to serve as both test subjects *and* operators. These types of experiments are particularly useful, as they help researchers better understand how the human body adapts to spending long periods of time in microgravity.

Defining the schedule to train astronauts to perform science is a complex process. Instructors must first determine how many crewmembers will be trained on each experiment, how many hours of training are required, and who will perform the training. Then, they must determine what procedures and software will be needed, and what equipment and facilities may be used, based on the budget available. Once all this information is established, individual training plans for each experiment are combined into a single plan that includes all the experiments in a particular scientific discipline.

Since crew time, whether before, during, or after flight, is a very limited resource, every detail of an experiment training session is planned, practiced, and coordinated. Often, due to the complexities of some experiments, the researcher or principal investigator (PI) instructs crewmembers in how to operate their experiment. Aiding the astronauts in their science training are Computer Based Training (CBT) lessons, developed by instructional design experts to provide ground-based and on-orbit crew training. CBTs are also useful for proficiency training when the crew is in orbit.

Exercise countermeasures

In between performing science, astronauts will spend a significant part of their day exercising, requiring them to conduct ground training on some of the exercise equipment they will be using while onboard the ISS.

Panel 7.8. Science onboard the International Space Station

One recent example of science conducted onboard the ISS for which astronauts had to be trained on the ground was the Anomalous Long Term Effects in Astronauts–Dosimetry (ALTEA-Dosi) experiment. Involving investigators from Italy and the US, the ALTEA-Dosi experiment was designed to assess the radiation environment inside the ISS's US Laboratory, *Destiny*.

Although astronauts are relatively well protected from radiation exposure in low Earth orbit (LEO), the effects of long-term radiation exposure are still poorly understood. With plans to send humans to the lunar surface for 6 months, where radiation levels will be much higher than in LEO, it is important radiation exposure be investigated more thoroughly. To that end, Italian and American scientists developed the ALTEA-Dosi experiment to measure the particle flux onboard the ISS in order to discriminate particle type, and measure particle trajectory and deposited energy. The experiment required astronauts to be trained in the operation of equipment such as the Intravehicular Charged Particle Directional Spectrometer (IVCPDS), the Extravehicular Charged Particle Directional Spectrometer (EVCPDS), and a helmet-shaped device holding six silicon particle detectors, designed to measure cosmic radiation passing through the detectors.

Astronauts responsible for administering the ALTEA-Dosi experiment were trained to position the helmet at a specific angle, before starting the test protocol from a laptop. At the beginning of each ALTEA-Dosi measurement, the astronaut had to start the automatic set-up/calibration procedure and the Dosi session. At the end of the session, the astronauts ensured the data collected by the particle detectors were sent to Earth in real time via automated telemetry.

Astronauts must spend so much time exercising because exposure to microgravity produces adaptations in nearly every physiological system. Some adaptations, such as motion sickness, are self-limiting, with symptom resolution occurring within days, while other adaptations produce more progressive changes, the most serious being those imposed upon the skeletal system, which shows no signs of resolution. One of the most regularly documented physiological changes associated with the spaceflight environment is the process of bone demineralization, caused by the absence of weight-bearing while in microgravity. An absence of load removes not only the direct compressive forces on the long bones and spine, but also the indirect loading on these bones from the pull of muscles on the various bone structures to which they are attached. Invariably, the unloading of the skeleton leads to osteoporosis, weakening of the bones, and delayed healing of fractures.

Exercise is one of a variety of countermeasures that have been incorporated into

Figure 7.20. Running on the International Space Station's treadmill. Image courtesy: NASA.

both short and long-duration space flights to help astronauts offset bone loss. During the Mir missions, cosmonauts regularly exercised for between 2 and 3 hr a day while being held with strong elastic cords against a running surface and supported by a belt around the waist – a practice that continues onboard the ISS today (Figure 7.20). However, despite various attempts to load the skeleton and all manner of exercise regimes, crewmembers continue to suffer bone loss and, after four decades of studying the effects of skeletal loading on bone growth, minimum loading thresholds are still unknown.

To offset the deleterious effects upon the cardiovascular, skeletal, and musculoskeletal systems, astronauts will spend at least 2 hr per day during their mission performing a broad program of exercise countermeasures. These exercises are designed to load the skeleton and induce mechanical strain upon the muscles. Crewmembers will follow an individual exercise training program developed preflight, based upon exercise testing, prior flight experience, and level of conditioning. Following each exercise session, physical trainers will receive a downlinked file containing heart rate and ergometer data of the crewmember's exercise training sessions. Every month, the crew will be subject to fitness assessments to determine aerobic capacity and strength levels. Based on the fitness assessments, recommendations and changes will be made to the exercise program to ensure countermeasures continue to be effective.

Medical training

In February, 2008, an undisclosed medical issue among the crew of the Space Shuttle *Atlantis* prompted a 24-hr delay to an EVA. ESA astronaut, Hans Schlegel, was eventually replaced by NASA astronaut, Stanley Love, and later rejoined the EVA rotation. The incident was typical of the many minor medical conditions astronauts suffer during spaceflight. To date, the spectrum of medical conditions reported by NASA and ESA astronauts have rarely required serious medical attention and there has been no medical evacuation of any NASA or ESA crewmember. However, given the extreme nature of the space environment combined with the extended duration of a typical expedition-class mission, it is inevitable that, sooner or later, medical intervention will be required to deal with one or more of the illnesses or injuries listed in Table 7.6.

Table 7.6. Classification of illnesses and injuries in spaceflight.[1]

Characteristics	*Examples*	*Type of response*
Class I • Mild symptoms • Minimum effect upon performance • Non-life-threatening	• Space motion sickness • Gastrointestinal distress • Urinary tract infection • Upper respiratory infection • Sinusitis	• Self-care • Administration of prescription and/or non-prescription medication
Class II • Moderate to pronounced symptoms • Significant effect upon performance • Potentially life-threatening	• Decompression sickness • Air embolism • Cardiac arrhythmia • Toxic substance exposure • Open/closed chest injury • Fracture • Laceration	• Immediate in-flight diagnosis and treatment • Possible evacuation • Possible mission termination
Class III • Immediate severe symptoms • Incapacitating • Unsurvivable if definitive care unavailable	• Explosive decompression • Overwhelming infection • Massive crush injury • Open brain injury • Severe radiation exposure	• Immediate evacuation following resuscitation and stabilization if necessary • Comfort measures applied

Crew medical training

Faced with the possibility of dealing with either a Class II or III medical contingency, it is likely crewmembers will perform most necessary operations autonomously, with limited support from the ground. The reason for the requirement of autonomous health care while on the ISS is due to the communication latency, which limits operational capabilities. Although there is only a 1-sec delay in communications to the ISS, station-based health care maintenance is being designed to be increasingly autonomous – an important consideration if the crew does not include a physician.

Since it is not certain every mission will have a physician-astronaut, the burden of any in-mission medical contingency will fall upon the shoulders of the crew's medical officer (CMO). At present, the CMO is a pilot or scientist with 34 hr of medical training, whereas other crewmembers receive only 17 hr of pre-flight medical training. However, given the extended missions to the ISS, crew medical training may be increased and astronauts selected for ISS missions will follow a schedule similar to the one outlined in Table 7.7.

Table 7.7. NASA medical training for International Space Station crewmembers[2]

Training session	*Crew*	*Time*	*Time prior to launch*
ISS space medicine overview	Entire crew	0.5 hr	18 months
Crew health care system (CHeCS) overview	Entire crew	2 hr	18 months
Cross-cultural factors	Entire crew	3 hr	18 months
Psychologic support familiarization	Entire crew	1 hr	18 months
Countermeasures system operations 1	Entire crew	2 hr	12 months
Countermeasures system operations 2	Entire crew	2 hr	12 months
Toxicology overview	Entire crew	2 hr	12 months
Environmental health system microbiology operations and interpretation	ECLSS	2 hr	12 months
Environmental health system water quality operations	ECLSS	2 hr	12 months
Environmental health system toxicology operations	ECLSS	2 hr	12 months
Environmental health system radiation operations	ECLSS	1.5 hr	12 months
Carbon dioxide exposure training	Entire crew	1 hr	12 months
Psychologic factors	Entire crew	1 hr	12 months
Dental procedures	CMOs	1 hr	8 months
ISS Medical diagnostics 1	CMOs	3 hr	8 months
ISS Medical diagnostics 2	CMOs	2 hr	8 months
ISS Medical therapeutics 1	CMOs	3 hr	8 months
ISS Medical therapeutics 2	CMOs	3 hr	6 months
Advanced cardiac life support (ACLS) equipment	CMOs	3 hr	6 months
ACLS pharmacology	CMOs	3 hr	4 months
ACLS protocols 1	CMOs	2 hr	4 months
ACLS protocols 2	CMOs	2 hr	4 months
Cardiopulmonary resuscitation	Entire crew	2 hr	4 months
Psychiatric issues	Entire crew	2 hr	4 months
Countermeasures system evaluation operations	CMOs	3 hr	4 months
Neurocognitive assessment software	Entire crew	1 hr	4 months

Table 7.7. *continued*

Training session	*Crew*	*Time*	*Time prior to launch*
Countermeasures system maintenance	Entire crew	2.5 hr	4 months
Environmental health system Preventive and Corrective Maintenance	Entire crew	1 hr	4 months
ACLS "megacode" practical exercise	Entire crew	3 hr	3 months
Psychologic factors 2	Entire crew	2 hr	1 months
Medical refresher	Entire crew	1 hr	2 weeks
CMO computer-based training	CMOs	1 hr/ month	During mission
CHeCS health maintenance system contingency drill	Entire crew	1 hr	During mission

BAROTRAUMA - EAR BLOCK/SINUS BLOCK
(ISS MED/3A - ALL/FIN) Page 1 of 1 page

BAROTRAUMA - EAR BLOCK, SINUS BLOCK

NOTE
Symptoms result from reduction in barometric pressure and expansion of trapped gas. Symptoms may occur during decompression preceding EVA or following loss of cabin pressure. Pain should resolve after repress in most cases. Persistent ear pain following repress requires examination.

Symptoms
Abdominal distention
Ear pain
Inability to clear ear
Loss of hearing acuity
Sinus pain
Toothache
Jaw pain

Treatment

AMP (blue)

1. If ear pain persists following repress, perform Otoscope Exam (Physical Exam-9).
   Look for the following signs:
   Red, inflamed ear drum
   Perforation of eardrum
   Drainage from ear drum, clear or bloody

2. Contact Surgeon with results.

Figure 7.21. Medical algorithm used by crews on the International Space Station. Image courtesy: NASA.

To ensure adequate treatment and rehabilitation during extended ISS missions, space agencies rely on instructing the crew with curricula and algorithms (Figure 7.21) based on microgravity physiological models of a human patient simulator (HPS). The curriculum includes medical training as well as telementoring and telemedicine techniques, based on the high-fidelity environment analog training (HEAT) concept that recreates a patient-care facility on the Moon.

Telemedicine

In the event of a serious injury requiring surgical intervention, it is probable remote or telepresence surgery – telemedicine – will be required (Panel 7.9 and Figure 7.22). Using this method of surgery, a surgeon in a remote location controls the robotic instruments performing the actual surgery. The method, which has been practiced successfully over intercontinental distances, has also been investigated by NASA, conducting research aboard its undersea research station, Aquarius, 20 m below the ocean, near the coast of Key Largo, Florida.

Panel 7.9. Telemedicine

A real experience with basic life support (BLS) and ALS in microgravity does not exist. Apart from surgery on experimental animals in orbit, emergency medical procedures have only been simulated in mockups, centrifuges, and in parabolic flight, but the inherent limits of these simulations are short, 20–25-sec microgravity periods, alternating with 2 G acceleration as the aircraft follows the parabolic flight path. Different methods for performing chest compressions have been tested, including CPR on a "free-floating" patient, but it is unlikely an efficient mechanical CPR can be performed over longer periods in microgravity or in partial gravity with presently recommended terrestrial methods. Airway management by conventional means (mask and bag, tracheal tube, laryngeal mask airway (LMA)) has been tested by several investigators; the average times to secure an airway in microgravity were longer in comparison with 1 G for both skilled and unskilled operators, but all procedures are technically possible.

NEEMO

> "We have learned that it is possible, and quite safe, to telementor an untrained person through a complex medical task."
>
> Dr Mehran Anvari, Principal Investigator,
> McMaster University Centre for Minimal Invasive Surgery

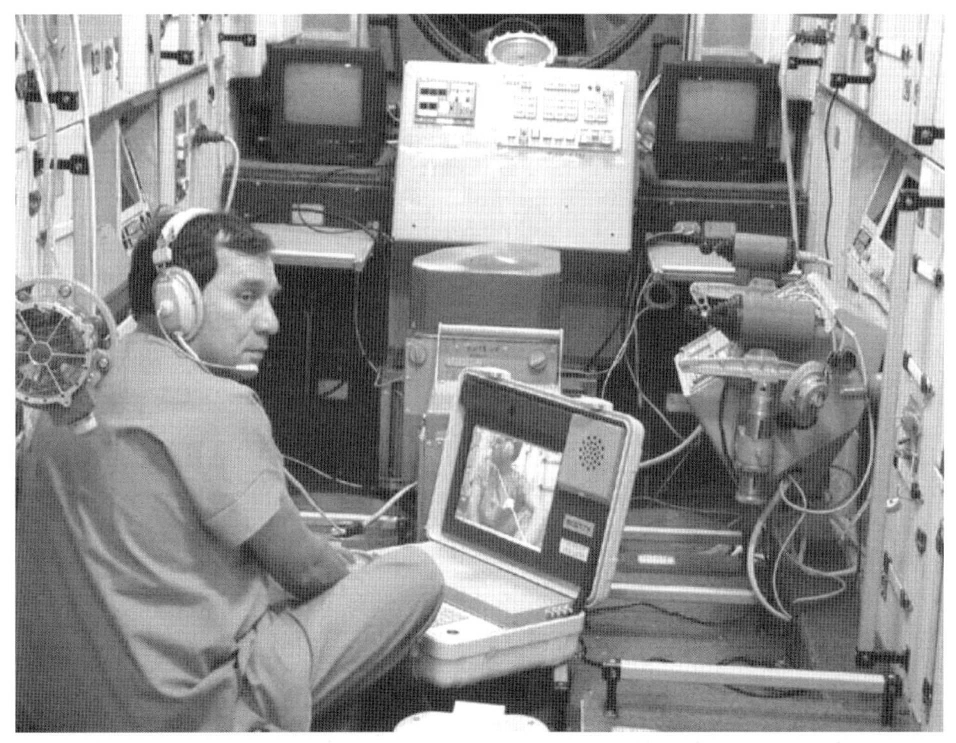

Figure 7.22. Testing different telemedicine architectures in the mockup of the ISS in the Gagarin Cosmonaut Training Center, Star City, Russia. Image courtesy: DLR.

During NASA's Extreme Environment Mission Operations (NEEMO) mission in October, 2004, Dr Anvari remotely guided the crew through gall bladder surgery and suturing of arteries while remaining in his hometown of Hamilton, Ontario. Although three of the six-member crew were physicians, none was a surgeon. In addition to the assistance provided by Dr Anvari, NEEMO's crew were aided by the Zeus system, a robot designed for the purpose of telemedicine.

The 2004 NEEMO mission and subsequent NEEMO missions have demonstrated telemedicine technology to be a tool capable of providing downlinks of video and diagnostic procedural images, in addition to connecting a crewmember with qualified surgeons. However, despite the sophistication of telemedicine, technology will never be a substitute for the skills of a well trained CMO and a motivated crew. Although it is possible to anticipate and prepare for common surgical and medical emergencies, unanticipated events do and can occur. For example, a subarachnoid hemorrhage was managed by the ingenuity and innovation of physicians in Antarctica, despite inadequate facilities.[3] Such an event not only underlines the human drive to save life against all odds, but also demonstrates how improvisation can lead to a successful outcome. Given the proven capabilities of telemedicine, it is likely it will become an increasingly common feature in astronaut training schedules, especially as a date for a return to the Moon looms closer.

The training increments described in this chapter cover most of the preparations an astronaut must complete during their IST phase. Mission training is intense and challenging work that is tackled with laser-like focus by astronauts whose primary goal is to ride a rocket into space. Upon completion of IST, they are ready to do just that. The next section describes the events leading to that eagerly anticipated goal, but, before following the astronauts during their final pre-launch preparation, it is appropriate, given NASA's and the ESA's far-reaching goals, to consider how astronaut training may change in the near future.

REFERENCES

1. Houtchens, B.A. Medical-Care Systems for Long-Duration Space Missions. *Clin. Chem.*, **39**(1), 13–21 (1993).
2. Jennings, R.T.; Sawin, C.F.; Barratt, M.R. Space Operations. In: R.L. DeHart, and J.R. Davis (eds), *Fundamentals of Aerospace Medicine*, 3rd edn. Lippincott Williams and Wilkins (2002).
3. Pardoe, R.A. Ruptured Intracranial Aneurysm in Antarctica. *Medical Journal of Australia*, **1**, 344–350 (1965).

8

Astronaut selection and training in the future

> "Men wanted for hazardous journey. Low wages, bitter cold, long hours of complete darkness. Safe return doubtful. Honour and recognition in event of success."
>
> Advertisement rumored to have been posted by Sir Ernest Shackleton before the launch of his legendary 1914 Imperial Trans-Antarctic Expedition

The advertisement placed by the great Shackleton may have been apocryphal but its content applies equally to those selected for future exploration class missions. The expeditions embarked upon by Shackleton, Fridtjof Nansen, and Douglas Mawson, almost a century ago, resemble in many ways the conditions of isolation and confinement that will be experienced by space travelers embarked upon interplanetary expeditions. The conditions will be different, but many of the problems confronting future space explorers will be the same ones that troubled explorers in the past – a reality reflected in some of the unique selection and training criteria described here.

FUTURE CREW SELECTION CRITERIA

If all goes to plan, a lunar outpost will be established some time after 2025. With this goal achieved, NASA and the European Space Agency (ESA) will set their sights on a manned mission to Mars. Even after 2025, it is likely a manned expedition to Mars (Figure 8.1) will lie on the ragged edge of achievability. Human exploration of this scope will require the optimum functioning not only of spacecraft, but also of astronauts. Failure of either could result in mission failure. Success therefore will depend not only on selecting (Table 8.1) the right group of astronauts, but also on training them effectively.

Figure 8.1. Plans to land astronauts on Mars will require new selection and training procedures. Pictured is a thermal nuclear rocket approaching Mars. Image courtesy: John Frassanito and Associates.

Table 8.1. Mars mission selection requirements.

Personal requirements	
Male	Age: >50
Meet space agency medical standards	Possess sense of community
Free of psychological problems	Possess effective conflict resolution skills
Technically competent	Possess sense of teamwork
Medical requirements	
Genetically screened for disease	Above average bone density
Undergone appendectomy	Genetically screened for radiation resistance
Undergone gall bladder removal	Screened for kidney stones
Social skills and behavioral traits	
Social compatibility	Tolerance
Emotional control	Agreeable and flexible
Patience	Practical and hard-working
Introverted but socially adept	Does not become bored easily
Sensitive to the needs of others	Desire for optimistic friends
High tolerance for lack of achievement	High tolerance for little mental stimulation
Self-confident without being egotistical	Subordination of own interests to team goals
Crew compatibility traits	
Tactfulness in interpersonal relations	Effective conflict resolution skills
Sense of humor	Ability to be easily entertained

Crew composition

Crew size

> "The human factor is three quarters of any expedition."
>
> Legendary Norwegian polar explorer, Roald Amundsen

As was the case with the crewmembers of Amundsen's and Shackleton's expeditions, the austere and isolated conditions facing interplanetary explorers will impose significant hardship. It will be assumed an astronaut has the skills and knowledge necessary to perform the duties of a crewmember, but these abilities will count for nothing if he or she cannot get along with others for several years in the confines of a vehicle/habitat no larger than a school bus! Given the unique characteristics of an exploration-class mission, the issues of crew composition and crew compatibility clearly become important selection factors due to the potential for these parameters to impact negatively upon a mission.

When NASA and/or the ESA finally embark upon a Mars mission, it is likely the size of the crew will fit with the current belief that smaller is better. Such a policy was implemented on many of the most successful polar expeditions, such as Nansen's 3-year Fram venture, which comprised just 13 crewmembers, and Shackleton's Imperial Antarctic Expedition, which consisted of just 27. Despite extreme isolation and prolonged confinement, Nansen's and Shackleton's expeditions were characterized by few interpersonal problems, thanks largely to small, homogeneous crews. This lesson is unlikely to be overlooked when it comes to defining the composition of Earth's first interplanetary crew.

Selecting crew roles

The occupational role of each member of an interplanetary crew has yet to be determined, but it is almost certain one crewmember will be a pilot and it is likely, given the extended duration of the mission, another crewmember will be a medical doctor. The role of commander will be assigned to the crewmember with the most experience and will not necessarily be the pilot, as has been the case in many space missions. Given the science objectives of such a mission, it is inevitable the crew will include at least one scientist, and other crewmembers will be extensively cross-trained in various scientific disciplines.

Crew gender

The issue of whether a crew should be all-male, all-female or mixed remains a contentious matter. Some have argued a female crew would exhibit preferable interpersonal dynamics and be more likely to choose non-confrontational approaches to solve interpersonal problems. Others have made a case for a mixed crew, claiming crews with women are characterized by less competition and seem to get along better. Evidence from Antarctic winter-over crews supports each of these arguments and suggests women, in addition to their mission function, would serve a socializing purpose. However, the introduction of a single female into a male group may have destabilizing effects because of sex issues – a topic that space agencies are notoriously reluctant to discuss! Fortunately, for mission planners at least, research indicates a drop in the sex hormone levels of astronauts during long-duration missions, resulting in a pronounced decline in sex drive, so the sex problem may not be an issue. However, just to be on the safe side, it is likely the first mission will comprise an all-male crew. Such a decision will be based not only on the psychosocial issues, but also because males are more radiation-resistant, thereby reducing the overall mission risk.

Crew compatibility

Determining crew compatibility has often been viewed as an opaque process. Perhaps the most diverse crew ever launched into orbit was Shuttle mission STS-51G, comprising civilian and military NASA astronauts, both male and female, a Saudi Arabian prince, and a French cosmonaut. Despite the mission's obvious multinational nature and cross-cultural challenges, the STS-51G crew was, by all accounts, a harmonious one that worked effectively throughout the mission. However, typical Space Shuttle missions last no more than 2 weeks, whereas an interplanetary mission will last 2 years or more. Crew compatibility issues will therefore obviously assume increasing significance in determining the effectiveness of the mission.

The problem in determining crew compatibility is that there is no one measure to predict whether a crew will work together effectively. Some researchers favor the use of psychological performance tests and personality questionnaires. Other investigators prefer a more behavior-oriented approach. The Russians, who have invested considerably in developing methods to assess interpersonal compatibility, consider biorhythms a useful tool for selecting cosmonauts! Perhaps the most useful lessons on the subject of crew compatibility can be found in research conducted during Antarctic winter-over increments and the annals of polar exploration. In fact, a veritable cornucopia of knowledge regarding crew composition and compatibility exists, thanks to the successful expeditions of Shackleton, Nansen, and Amundsen, and experience from Antarctic research stations. This history of crew dynamics in harsh environments provides mission planners with more than enough information to carefully select a compatible crew.

Exceptional selection criteria

Due to the unique characteristics of what may be a 3-year mission, space agencies will employ some rather unconventional selection criteria, such as genetic screening and precautionary surgery. Given the unusual nature of these criteria, those selected will cross a legally defined boundary; in the same way as a soldier relinquishes certain individual rights when joining the military, astronauts chosen for a Mars mission will be expected to do the same and accept collective standards contributing to the common good of realizing a successful mission.

Genetic screening

Current international legislation bars employers from using the genetic information of individuals when making hiring decisions. However, space agencies tasked with the onerous task of selecting perhaps the most unique space crew in history will surely be exempted from this requirement.

Rationale

Prospective Mars astronauts, just like Ethan Hawke's character in *Gattaca* (Figure 8.2), will likely be genetically tested during the medical examination of the astronaut recruitment campaign. Genetic testing will allow the space agencies to diagnose vulnerabilities to inherited diseases that may cause problems during long missions. The testing will also reveal information concerning the presence of genetic diseases

Figure 8.2. Future crew selection policies may echo those portrayed in the film *Gattaca*. Image courtesy: IMDB.

and mutant forms of genes associated with increased risk of developing genetic disorders. Additionally, genetic testing can confirm or deny a suspected genetic condition and provide information concerning the possibility an astronaut may develop a disorder.

Types of testing

Diagnostic testing will be used to eliminate most genetic or chromosomal conditions, while *carrier* testing will be used to identify candidates carrying one copy of a gene mutation that may cause a genetic disorder. *Predictive* testing will be used to detect gene mutations associated with disorders present in the candidate but where no features are present at the time of testing (Panel 8.1). This type of testing will identify those at risk of developing a disease, such as cancer, during a mission. Obviously, if any of the results are positive, the candidate will be eliminated from the recruitment process.

Panel 8.1. Types of testing explained

Diagnostic testing is used to identify or rule out a specific genetic or chromosomal condition. In many cases, genetic testing is used to confirm a diagnosis when a particular condition is suspected based on physical signs and symptoms. Carrier testing is used to identify people who carry one copy of a gene mutation that, when present in two copies, causes a genetic disorder. Predictive testing is used to detect gene mutations associated with genetic disorders that appear after birth, often later in life. These tests can be helpful to people who have a family member with a genetic disorder but who have no features of the disorder at the time of testing.

The tests described are performed on a sample of blood, hair, or skin, which is sent to a laboratory, where technicians search for differences in chromosomes, DNA, or proteins. Due to the problems in interpreting genetic tests, space agencies will need to exercise particular care in determining the genetic profile of candidates. For example, a negative test result means the laboratory did *not* detect an abnormal gene, chromosome, or protein. However, although such a result may indicate a person is not affected by a particular disorder, it is possible the test missed a disease-causing genetic alteration. This is because some tests simply cannot detect *all* genetic changes associated with a specific disorder. To eliminate any ambiguity, space agencies will hopefully discard uninformative and inconclusive tests and conduct secondary tests. However, in the event of a positive result, the likely consequence for the candidate will be elimination from consideration as an astronaut. Needless to say, the effect of a positive result on candidates who have spent their professional lives accumulating the qualifications to become an astronaut will be devastating. However, such testing will be necessary, given the potentially dire consequences of an astronaut being diagnosed with a critical illness during the mission.

Precautionary surgery

Once a crewmember is pronounced genetically free of any future disease or disorder and is provisionally selected for the Mars mission, they may be required to undergo precautionary surgery. While a candidate may be required to undergo a number of precautionary procedures, the most likely one is removal of the appendix.

Appendicitis

The appendix is a closed-end narrow tube attached to the first part of the colon. If the opening to the appendix becomes blocked or the fatty tissue in the appendix swells, bacteria, normally found within the appendix, may invade and infect the wall of the appendix. This infection results in *appendicitis*, to which the body responds by inflaming the appendix, which may ultimately lead to rupture, followed by spread of bacteria outside the appendix. Alternatively, the appendix may become perforated leading to an abscess or, in some cases, the entire lining of the stomach may be infected. Appendicitis may require the contents of the stomach to be drained through a tube passed through the nose. Needless to say, in the confined environment of a spacecraft in zero-gravity, such a procedure would challenge even the most experienced surgeon! Perhaps the most feared complication of appendicitis is *sepsis*, a condition in which bacteria enter the blood and infect other parts of the body. Even on Earth, sepsis is considered a serious complication, but to an astronaut bound for Mars or returning to Earth, such a complication would be a death sentence.

Medical support

These complications alone represent a powerful argument for removing the appendices of Mars-bound crewmembers, but there are other factors to consider, such as diagnosing the condition – a procedure that would use vital medical consumables. For a crewmember suspected of suffering from appendicitis, the only possible diagnostic procedures available would be a urinalysis and an ultrasound procedure. In the event of complications, a computer tomography (CT) scan and abdominal X-ray would be unavailable due to the limited medical resources, although a laparoscopy could possibly be performed. However, a laparoscopy – a procedure in which a small fiberoptic tube with a camera is inserted into the abdomen through a puncture hole made in the wall of the stomach – requires a general anesthetic and would present a challenging procedure in zero-gravity.

Furthermore, even on Earth, appendicitis is often difficult to diagnose because other inflammatory problems can mimic the symptoms of the condition.

Appendectomy

Should a crewmember be correctly diagnosed with appendicitis, the next problem would be treatment, involving the removal of the appendix in a procedure known as an *appendectomy*. This requires the surgeon to make a 4–6-cm incision in the skin and layers of the abdominal wall, in the area of the appendix, and to remove the appendix. If an abscess is present, pus must be drained before the abdominal incision is closed. In recent years, laparoscopic surgery has been used to perform the procedure but in zero-gravity, the method would present risks.

Once the candidate Mars crewmember has been genetically screened, and has undergone the requisite precautionary surgery, they will undoubtedly breathe a sigh of relief and look forward to mission training! However, for those with their sights set on Mars, the medical aspects of preparation will still not be over.

Cryopreservation

It is known short-duration spaceflight has no adverse effect on the ability of astronauts to conceive and bear healthy children. However, pregnant astronauts are not allowed to train in the vacuum chamber, KC-135 Zero-G aircraft, fly in T-38s, or train in the Neutral Buoyancy Facility (NBF). Due to the training requirements for a Mars mission, any astronaut becoming pregnant would likely be dropped from their flight assignment. Another problem for astronauts wishing to conceive is the effect of a long-duration mission upon fertility and the increased chance of genetic defects as a result of prolonged exposure to deep space radiation. The effects of a return trip to Mars upon the reproductive capacity of humans have yet to be characterized, so it is difficult for mission planners to provide guidance concerning the risks of radiation. The obvious solution will be to implement a cryopreservation program. For women, *banking* (of embryos) will eliminate the potential problems of damage to embryos by galactic cosmic radiation (GCR) and solar particle events (SPEs). It will also augment fertility, since the pregnancy and miscarriage rates for embryo transfer are dependent on the age of the embryos at the time of collection. Another option open to female astronauts will be to cryopreserve ovarian tissue. For male astronauts, sperm cryopreservation will be implemented as an option in the very likely event of their returning to Earth infertile.

FUTURE CREW TRAINING

Bioethics

Current ethical standards for the selection and training of astronauts were developed in an era of short-duration space missions when repeat missions were the norm and a return to Earth within days was possible. In future missions beyond Earth orbit, a diverse group of astronauts will venture to remote destinations for increasingly long periods. Contact with Mission Control will be delayed and rapid return impossible. Long-duration missions to the Moon and multi-year missions to Mars will inevitably create special circumstances (Table 8.2) for which current ethical standards governing the selection and training of astronauts are inadequate. As the prospect of multi-year missions approaches, mission planners will design a new ethical framework to guide mission commanders and crewmembers in their decision making when it comes to dealing with some of the potentially awkward moral questions.

Table 8.2. Bioethical selection and training issues for long-duration missions.

How do you get rid of the dead body of a crewmember?
When should life support be discontinued for a critically ill astronaut consuming valuable oxygen?
Should a straitjacket be included in the medical supplies?
Should NASA mandate prophylactic surgery such as removal of appendix, tonsils and gall bladder before a mission to Mars?
If a crewmember becomes disabled during the mission, who becomes their surrogate decision maker? Their spouse? NASA physicians? Other crewmembers?
Should NASA mandate genetic screening as a part of the astronaut recruitment process?

Sex

Perhaps the most discussed ethical quandary is the issue of how to cope with sexual desire among a crew of healthy young men and women. While sex has long been an almost taboo topic among mission planners, the question of how to address sexual desire is perhaps the easiest to solve by simply selecting an all-male or all-female crew. Although such a policy may deselect better qualified candidates, mission planners may judge that the behavioral issues sex poses on a multi-year mission outweighs such disadvantages. Some may argue an all-male crew would create its own problems. However, the precedent set by innumerable multi-year polar expeditions, such as Shackleton's Imperial Trans-Antarctic Expedition (Figure 8.3), provides overwhelming evidence that spending time in a confined environment

Figure 8.3. Sir Ernest Shackleton. Image courtesy: NOAA.

with a group of men for several years under extreme duress does not result in any mission-compromising behavioral issues.

Terminal illness

More pressing ethical questions concern what action to take if an astronaut becomes terminally ill during a mission. In such an event, the Commander may be directed by Mission Control to euthanize the ill crewmember in order to preserve medical supplies and life-support consumables. Alternatively, the affected crewmember, knowing he/she has only a short time to live, may offer to sacrifice his/her life for the mission. In such a situation, what will mission guidelines instruct the Commander to do? Needless to say, euthanizing a crewmember will not endear a space agency to the media or the public, who assume an astronaut's well-being will take precedence over mission success. Such a perception is not surprising, since, to date, any astronaut becoming ill or injured onboard the ISS has had the opportunity to simply leave the outpost onboard the Soyuz and return to Earth within hours. Unfortunately, this will not be possible when the nearest hospital is several million kilometers away! For situations such as "who gets thrown from the lifeboat", it will be necessary to equip mission commanders and crew with the necessary ethical framework to make difficult decisions.

Hibernation

Another important element of training will be a theoretical and practical familiarization with the process of hibernation. While this procedure is currently barely beyond the science fiction arena, the technology may be operational in time for the first Mars mission. If thoughts of long-duration space journeys and hibernation conjure up images of the opening scene of *Alien*, you're not alone. However, the technology exists no longer merely in the realm of science fiction movies, as NASA and the ESA are funding research into methods enabling astronauts to spend months in a state of suspended animation. Although the concept may seem futuristic, the daunting timeframe facing astronauts means hibernation is an idea to be taken seriously. Apart from the boredom of a lengthy transit, there are powerful logistical reasons to place astronauts in hibernation. Both the ESA and NASA have estimated a typical return trip to Mars would require 30 tonnes of consumables for a crew of six. In addition to the food issue, there are the not inconsiderable matters of waste generation and oxygen consumption. Hibernating astronauts will require less oxygen and food than active astronauts, thereby resulting in a lighter spacecraft and less fuel (Table 8.3)! Already, scientists and engineers are designing "sleep pods" (Figure 8.4) that may resemble those on the *Nostromo*, the spacecraft in the film *Alien*.

Table 8.3. Effect of hibernation on life support requirements.[1]

Life support component	*Purpose*	*Effect of hibernation*
Atmosphere management	Air revitalization, temperature, humidity and pressure control Atmosphere regeneration Contamination control	Reduced heating requirement Reduced regeneration requirement
Water management	Provision of potable and hygienic water Recovery and processing of waste water	Reduced significantly
Food storage	Provision of food	Reduced significantly
Waste management	Collection, storage and processing of human waste	Reduced significantly
Crew safety	Fire detection and suppression Radiation warning system	Augmented systems required
Crew psychology	Maintenance of crew mental health	Reduced
Crew health	Bone demineralization and muscle atrophy	Augmented systems required

Figure 8.4. Astronauts hibernating on their way to Mars may sleep in hibernaculums similar to the ones portrayed in the classic sci-fi film, *Alien*. Image courtesy: IMDB.

Figure 8.5. California Ground Squirrel (*Spermophilus beecheyi*). Image courtesy: Free Software Foundation/Wikipedia.

Will hibernation actually work? Scientists working for the ESA's Advanced Concepts Team (ACT) seem to think so. The ACT scientists have studied other species, such as bears, ground squirrels, and rodents, for which hibernation (Panel 8.2) is a regular part of their lives. Already, researchers have been able to chemically induce a stasis-like state in living cells and have progressed to small, non-hibernating mammals like squirrels[2] (Figure 8.5).

Panel 8.2. Hibernation science

The key to putting astronauts in a state of hibernation may lie in a synthetic, opioid-like compound called *Dadle*, or *Ala-(D) Leuenkephalin*, which, when injected into squirrels, can put them in a state of hibernation during the summer.[2] This research has been already extended to studies investigating the effect of applying Dadle to cultures of human cells, revealing human cells divide more slowly when Dadle is applied. In conjunction with the studies investigating Dadle, researchers are testing compounds such as *dobutamine* and *insulin-growth factor* (IGF). Dobutamine is normally administered to bedridden patients to strengthen their heart muscles but, in the case of hibernating astronauts, the compound would be administered to maintain health during the long period of inactivity. IGF would be administered to boost the astronauts' immune systems, which would be depressed during the long period of inactivity.

Training for hibernation

How would the process work? First, crewmembers would be required to attain a very high level of fitness to maximize their body's ability to deal with the stress of hibernating and the deleterious effects of being in a state of suspended animation for several months. Astronauts would then be subjected to short hibernation increments of 7–10 days. These hibernation indoctrination sessions would take place in a *hibernaculum*, a highly advanced medical facility. Here, flight surgeons would connect astronauts to intravenous tubes, through which fluids and electrolytes would be administered to compensate for changes in blood composition during the hibernation. Then, administration of a hibernation-inducing compound would place the astronauts in a state of hibernation. During the hibernation, a suite of medical sensing and hibernation administration facilities would monitor the state of the hibernating astronauts. In addition to ensuring body temperature, heart rate, brain activity and respiration stay within normal boundaries, the medical equipment would also monitor blood pressure, blood glucose levels, and blood gases. After their training hibernation session, astronauts would be woken and be assessed by the flight surgeons. Awakening from a state of deep torpor will be a valuable training experience due to the potential of undetermined physiological and behavioral outcomes (Panel 8.3).

Panel 8.3. Behavioral effects of hibernation

Research has shown the deep torpor associated with hibernation may be problematic for the brain.[3, 4] This finding is a concern for those on their way to Mars – a mission requiring a fully functional and intact brain! The problem occurs during torpor entry, when the body's temperature is gradually reduced. The cooling process results in reduced cortical power and profound differences in sleep architecture and memory consolidation. More worrying for those awakening from hibernation are the potentially deleterious effects upon spatial memory and operational conditioning. Of course, until hibernation is performed on humans, we just won't know for sure.

In the event of a contingency during an actual mission, the hibernation suite would be fitted with in-situ monitoring by an artificial intelligence (AI) agent operating on principles similar to medical monitoring systems. The agent would monitor the hibernation medical equipment, ensure environmental parameters were maintained, and monitor contingency events, such as solar storms or a mid-course correction that might require waking the crew.

Although placing astronauts into hibernation would solve many problems during the deep space phases of a Mars mission, several issues remain unresolved. Scientists still need to develop a trigger compound capable of inducing a state of hibernation and research concerning the secondary effects of hibernation is still lacking. For example, the effects of hibernation on memory, the metabolism, or the immune system are unknown.[4] Another problem are the deleterious effects of zero-gravity combined with the inactivity of hibernation, although this may be resolved by using some means of artificial gravity. Other challenges include problems associated with how the hibernation state is induced, established, regulated, and exited, and how to administrate compounds to a hibernating human. Achieving and perfecting human hibernation will require expertise in, and integration of, pharmacology, genetic engineering, environmental control, medical monitoring, AI, radiation shielding, therapeutics, spacecraft engineering, and life support. Only when *all* these disciplines have been successfully integrated will human hibernation be able to make long-haul spaceflight a little more comfortable.

Advanced surface exploration training

Surface activities conducted by the first Mars crews will require astronauts to be trained in a new set of skills. The NASA reference mission to Mars calls for a two-phased approach. In phase 1, astronauts, supported by robotic systems, will explore the Martian surface, collect and analyze geologic and meteorological data, search for base sites, and conduct technology verification experiments. In phase 2, a Mars site will be selected and a permanent base will be constructed (Figure 8.6). Achieving the objectives of the two phases will require astronauts to operate in-situ resource utilization (ISRU) equipment, nuclear power systems, a greenhouse, erect inflatable habitats (Figure 8.7) and laboratories, and build structures inside the habitats. Once the base has been established, astronauts will turn their attention to surface exploration activities, requiring the utilization of rovers, crawlers, blimps, air drills, and biomimetic explorers.[5–8] Testing the feasibility of these systems and exploration architectures will require astronauts to spend time in analog environments such as the Haughton Crater on Devon Island. Here, astronauts bound for Mars will test the integrity of the Mars habitat, verify communication architectures, and validate revolutionary exploration architectures.

Bioinspired engineering of exploration systems

A manned Mars mission will be a sophisticated expedition, demanding the greatest complexity and functionality of any mission ever undertaken. Although astronauts will spend a large proportion of their time conducting exploration activities, a significant amount of their workday will also be spent attending to life support systems, repairing equipment, and performing routine maintenance tasks. Ideally, then, to ensure the science return is maximized, it would be helpful if the astronauts could deploy autonomous explorers. These systems would be capable of surveying

Figure 8.6. A future Mars base. Image courtesy: NASA.

Figure 8.7. An inflatable habitat that may be used to construct a Mars base. Image courtesy: NOAA.

the surface of Mars, conducting science experiments, and performing scouting missions. Fortunately, the means to achieve this goal are available thanks to a multidisciplinary concept known as bioinspired engineering of exploration systems (BEES).

BEES represents a new approach NASA and ESA are adopting to space exploration. By studying how living things interact with their environment, the agencies plan to apply similar principles to explore planets such as Mars. However, the intent is not just to mimic operational mechanisms found in a specific biological organism, but to take the best features from a variety of diverse bio-organisms and apply them to a desired exploration function. Adopting this approach, engineers hope to build explorer systems endowed with capabilities exceeding those found in nature, as the systems will possess a combination of the very best nature-tested mechanisms for a specific function. Such an approach is a logical one, since, through billions of years of evolution, nature has perfected its designs and, by selecting the best of these designs, it will be possible to go beyond biology and achieve unprecedented capability and adaptability in explorer systems.

Biomimetic explorers

Already, work is underway to develop what engineers call *biomorphic explorers*, using principles of a BEES discipline known as *biomimetics*. Biomimetics[8–10] deals systematically with the technical execution and implementation of construction processes and developmental principles of biological systems. Using biomimetic principles, engineers plan to develop the classification of biomorphic explorers into two main classes of surface and aerial explorers. Utilizing these autonomous explorers within a surface exploration architecture will require astronauts to be trained in the operation of biomimetic-inspired systems such as entomopters and gecko-tech.

Entomopters

During the 1990s, the Defense Advanced Research Project (DARPA) considered the feasibility of using minute flying vehicles, on the scale of insects. DARPA's research gave birth to the *Entomopter*, a descriptor combining the concept of *entom*ology with the word heli*copter*. It wasn't long before NASA, an agency familiar with the potential of airborne technologies to conduct exploration, decided to pursue its own research.

One of the problems of flying on Mars is the rarefied atmosphere. This means an aircraft must have a large surface area and fly at very high speeds to generate sufficient lift, which isn't very helpful for exploring. Another means of achieving the required lift is to flap the wings, but simply flapping up and down is still not sufficient to achieve flight in the Mars environment. To produce sufficient lift, the flapping must be combined with an additional lift-producing mechanism known as the Magnus force, a rotational motion insects use when flapping their wings. By replicating these insect flying mechanisms, researchers developed the Entomopter, capable of flight in the Martian environment and able to operate close to the surface.

Biomimetic training in an analog environment

Astronauts will conduct biomimetic training in analog environments such as Haughton Crater (Figure 8.8) on Devon Island. Here, astronauts will learn how to use the Entomopter to scout out potential locations of interest and conduct surveying tasks. They will also learn the basics of a baseline Entomopter mission, which will begin with crewmembers programming the day's tasks into the Entomopter's computer and ensuring the airborne refueling rover is operational. Once programmed, the astronauts will launch the autonomous explorer from its refueling rover and the Entomopter will begin its tasks. Although the Entomopter's flights will only last between 5 and 10 min, the flights will be numerous thanks to the refueling rover. To avoid crewmembers having to monitor the Entomopter while it surveys Haughton Crater, it will be fitted with flight control, navigation, and collision avoidance software, just like a regular aircraft. A rover-centric system will also be a part of the Entomopter's software, providing updates of its relative position to the refueling rover so it doesn't run out of fuel. Also, in common with regular aircraft, the Entomopter will be fitted with radar, enabling it to not only map

Figure 8.8. Haughton Crater on Devon Island, Canada. Image courtesy: Mars Society.

obstacles, but also provide it with an inherent homing beacon capability and bidirectional communication via direct transmission from the refueling rover.

Biomorphic missions

Once the astronauts have deployed the Entomopter, they may turn their attention to a small fleet of biomorphic explorers.[8, 10] Due to the explorer's versatility, the astronauts may assign a variety of science objectives in a series of biomorphic missions. These missions will be characterized by synergistic use of existing and conventional surface and aerial assets, such as the habitat and rover. Science objectives may include close-up imaging for identifying hazards, assessing geological sites, gathering atmospheric information, and deploying surface payloads such as surface experiments. To achieve these objectives, the reconfigurable biomorphic explorers will be programmed for specific functions.

Sample biomorphic missions: imaging and site selection

One of the first exploration objectives of the crew training in Haughton will be to survey and select sites of interest. During an actual mission, imaging from orbiting vehicles will allow only broad coverage with a spatial resolution limited to perhaps only 1 m. With such limited resolution, it will be difficult for astronauts to plan routes and characterize sample return sites. To achieve high-resolution coverage, astronauts will deploy bioflyers on a variety of flight paths to image the horizon terrain. Bioflyers will be equipped with a suite of sensors and miniature cameras. If a bioflyer identifies a potential exobiological site, it will decide to terminate the flight and deploy a small science experiment with a pyrotechnic device that will disengage from the bioflyer before penetrating the ground. Meanwhile, the bioflyer will continue orbiting the exobiological site waiting for data to be transmitted by the science payload. Once the data are received, the bioflyer will transmit the results to the habitat and, if required, continue to orbit the site to serve as a beacon for the crew. The crew will then begin *their* mission and drive to the site identified by the bioflyer. Thanks to the imaging information sent by the bioflyer, the crew will be able to take the most direct and safest route to the site, thereby saving time and reducing risk.

Sample biomorphic missions: surface experiments

In this training mission, astronauts will deploy biomorphic orbiters to an area of geological or scientific interest in the vicinity of Haughton Crater. The orbiters will carry seed wing flyers, each equipped with a small surface probe, chemical experiment payload, and a miniature camera. The orbiter will traverse an area of interest while simultaneously gathering meteorological data such as weather patterns used to select the timing of release of the seed wing flyers. Once an area of interest is identified, seed wing flyers will be released and land on the surface, where they will conduct surface experiments that may include testing for trace elements. Once the science experiment is complete, the seed wing flyer will transmit its data to the orbiter, which in turn will transmit the data to the astronauts, who will analyze it.

Sample biomorphic missions: aerial reconnaissance

There may be occasions when the orbiter information relayed to the astronauts is not sufficiently detailed to make an assessment of large-scale geological or meteorological data. In such an event, the astronauts may release a squadron of small biomorphic gliders equipped with small infrared (IR) cameras and surface probes. The gliders will be pre-programmed by the astronauts with coordinates provided by the orbiter and deployed to priority targets. As the gliders fly over their targets, they will transmit high-resolution imagery back to the astronauts. From their training habitat in Haughton Crater, the astronauts will be able to command the gliders to take different flight paths in order to image the area from different aspect angles and also instruct the gliders to land if a particular area appears it may yield valuable data.

Sample biomorphic missions: local and regional sample return

In this mission, the objective will be to obtain samples from potential exobiology sites and areas of geological interest. To do this, the astronauts will deploy an autonomous rover loaded with an arsenal of scientific experiments and a squadron of biomorphic explorers equipped with a miniature camera and a small IR detector.

At the target location, the rover will deploy the scientific experiments and transmit the results to the astronauts. If results appear interesting, the astronauts will command the biomorphic explorers to return samples while the rover continues on its mission.

Yabbies

Another biomimetic application astronauts may be trained in the use of while in Haughton is an engineered version of the common freshwater crayfish, known in Australia as the *Yabby* (Panel 8.4). At the University of Melbourne, scientists and zoologists think crews on Mars will be able to deploy teams of *robo-yabbies* to help search for water or conduct chemical analysis of the planet's crust – tasks that may be impractical or dangerous for humans.

> **Panel 8.4. The Yabby**
>
> Thanks to advances in computational network modeling (CNM), engineers have managed to apply the movement of the Yabby's tail to the design of robots capable of traversing over difficult terrain. The Yabby's segmented tail, which acts like a hinged lever capable of changing to act as a sail for steering or as a paddle for swimming, gave scientists the idea of using a similar system in the design of miniature, lightweight robots with multi-jointed legs, capable of performing a range of complex tasks needed to explore Mars.

The biomorphic explorers and Yabbies represent just a snapshot of the types of robots astronauts will be trained in the operation of before embarking on a Mars mission. However, the discipline of biomimetics isn't just limited to autonomous explorers. Astronauts training in analog environments such as Haughton may also use the terrain to practice using other biomimetic applications.

Gecko-Tech

Thanks to a mysterious gecko adhesive, geckos (Figure 8.9) have the ability to walk up walls and run across ceilings (Panel 8.5). These skills could also be usefully employed by astronauts exploring Haughton and the surface of Mars! In between practicing the operation of fleets of biomorphic explorers, astronauts may venture outside their analog training environment wearing a Gecko-Tech biomimetic suit.

Gecko-derived spacesuits would enable astronauts on Mars to scale the walls of the Martian valleys and explore almost any terrain by being able to climb over any obstacles, thereby saving time over a more conservative "going around" route.

Figure 8.9. The gecko may hold the key to spacesuits capable of scaling walls. Image courtesy: Wikipedia.

Panel 8.5. Gecko-Tech

The key to the gecko's mobility is achieved thanks to evenly spaced gripping strips crossing the end of the gecko's toe. These strips, known as *lamellae*, comprise special hair-like structures called *setae*. On each foot, there are about 500,000 setae, each seta consisting of tiny pad-on-stem structures called *spatulae*, which act as the gripping structure enabling the gecko to perform its acrobatic mobility. Research investigating materials possessing the necessary mechanical tensile strength, flexibility, and formability suitable for fabricating artificial spatulae and setae suggest gecko-tech adhesion is a realizable goal using nanometre-scale fabrication techniques.

Using the gecko-suit, designed with gecko-style adhesive pads on the hands, knees, and feet, Martian explorers would not only be able to climb vertically, but also achieve inverted mobility!

Artificial gravity indoctrination

> "An artificial gravity system was deemed necessary for the MSM's outbound hab flight to (1) minimize bone loss and other effects of freefall; (2) reduce the shock of deceleration during Mars aerobraking, and (3) have optimal crew capabilities immediately upon Mars landing. Experience with astronauts and cosmonauts who spent many months on Mir suggests that if the crew is not provided with artificial gravity on the way to Mars, they will arrive on another planet physically weak. This is obviously not desirable. Unless a set of countermeasures that can reduce physiological degradation in microgravity to acceptable levels is developed, the only real alternatives to a vehicle that spins for artificial gravity are futuristic spacecraft that can accelerate (and then decelerate) fast enough to reach Mars in weeks, not months."
>
> Mars Society Mission[11]

To realize the goal of a human mission to Mars, it will be necessary to mitigate the human risks associated with prolonged weightlessness. Despite the experience of long-duration missions onboard the International Space Station (ISS), no completely effective countermeasure, or combination of countermeasures, exists. In fact, the operational countermeasures currently employed have failed to fully protect astronauts for more than 3 months in low Earth orbit (LEO). It seems unlikely, therefore, that current countermeasures will protect astronauts venturing to Mars and back over a 30-month duration!

One solution is artificial gravity (Figure 8.10). While it may sound like science fiction (fans of *2001: A Space Odyssey* will recall artificial gravity was a feature

Figure 8.10. Research using NASA's short-radius centrifuge may hold the key to realizing artificial gravity onboard future spacecraft. Image courtesy: Wyle.

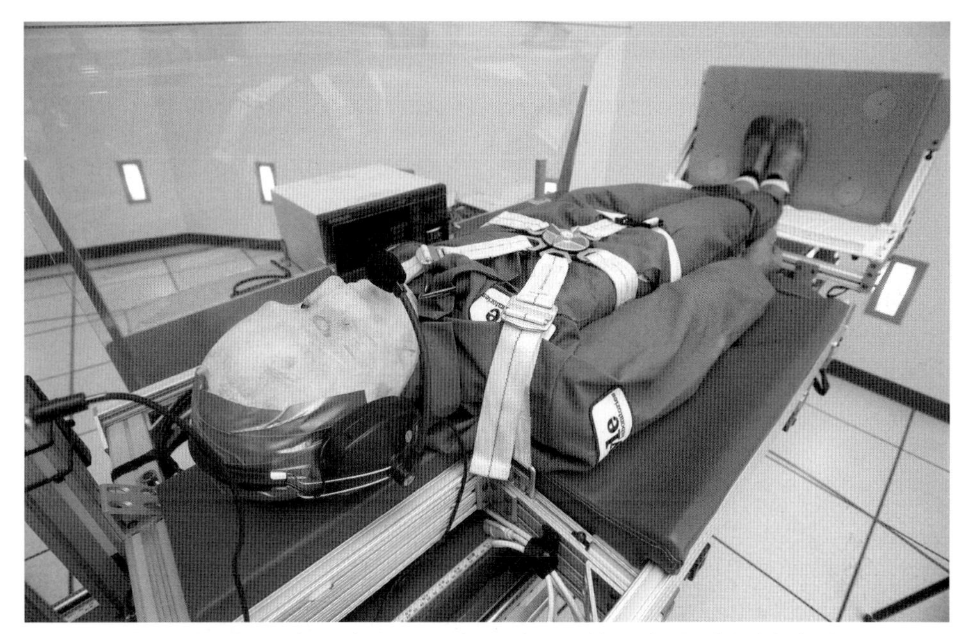

Figure 8.11. Configuration of mannequin and centrifuge in Wyle's lab in Houston. Image courtesy: Wyle.

onboard the film's spacecraft, *Discovery*), artificial gravity is a countermeasure that may maintain neurovestibular, cardiovascular, and musculoskeletal function for the duration of a trip to Mars. While it may not be a panacea for addressing *all* the risks associated with long-duration missions, artificial gravity offers promise as an effective countermeasure against physiological deconditioning effects.[12]

Unlike the rotating vehicles featured in many science fiction movies, the rotating torus is neither feasible nor necessary. The artificial gravity system onboard the Mars spacecraft will comprise a 3-m radius centrifuge (Figure 8.11). Astronauts will lie on a "subject station" or narrow bed-like sections that can be moved closer or farther from the short-radius centrifuge's (SRC) pivot point. As the centrifuge spins about the pivot, centrifugal forces will create G-loads along the subject's body axis (head to feet) proportional to the rate of rotation. During their artificial gravity indoctrination training, astronauts will be instrumented with heart-monitoring sensors, a pulse-measuring device, a blood-pressure cuff, and an oximeter. They will also wear a headset and boom microphone, enabling communication with the SRC control room. They will then lie horizontally on one arm of the SRC and will be strapped to the subject station using a five-point harness. A flat-panel screen directly above the astronaut will allow them to watch movies (or perhaps review technical documents describing how Entomopters work!) during the spin. The objective of the artificial gravity indoctrination will be to pre-adapt astronauts to a rotating environment and also to determine optimal centrifugation speeds and durations before the mission.

During the manned spaceflight era, much has been learned about astronaut selection and training. Nevertheless, the unique challenges of interplanetary missions will demand a re-evaluation and recalibration of the selection and training guidelines described in this book. These new guidelines will prepare astronauts to cope not only with the hostile environment of space, but also with the daunting challenges posed by extended isolation and confinement. In due course, given the unique selection and training requirements for such missions, it is likely it will be a new breed of astronaut who embarks upon multi-year missions to Mars and beyond. While they may not suffer the low wages and bitter cold endured by Shackleton's men more than a century ago, honor and recognition are almost guaranteed, in much the same way as they are for today's spacefarers.

REFERENCES

1. Hypometabolic Stasis in Astronauts for Long Term Space Flight. Insights from Fundamental Research 55th International Astronautical Congress of the International Astronautical Federation, the International Academy of Astronautics, and the International Institute of Space Law, Vancouver, British Columbia, October 4–8, 2004.
2. Wang, L.C.H. Time Patterns and Metabolic Rates of Natural Torpor in the Richardson's Ground Squirrel. *Canadian Journal of Zoology*, **57**, 149–155 (1979).
3. Millesi, E.; Prossinger, H.; Dittami, J.P.; Fieder, M. Hibernation Effects on Memory in European Ground Squirrels. *Journal of Biological Rhythms*, **16**, 264–271 (2001).
4. Strijkstra, A.M. Good and Bad in the Hibernating Brain. *Journal of the British Interplanetary Society*, **59**, 119–123 (2006).
5. Menon, C.; Ayre, M.; Ellery, A. Biomimetics: A New Approach for Space Systems Design. *ESA Bulletin*, **125**, 20–26 (2006).
6. Menon, C.; Broschart, M.; Lan, N. Biomimetics and Robotics for Space Applications: Challenges and Emerging Technologies. IEEE International Conference on Biomimetic Robotics, Rome, Italy, April 10–14, 2007.
7. Mjolsness, E.; Tavormina, A. The Synergy of Biology, Intelligent Systems and Space Exploration. *IEEE Intelligent Systems*, **15**(2), 20–25 (2000).
8. Thakoor, S.; Miralles, C.; Martin, T.; Kahn, R.; Zurek, R. *Cooperative Mission Concepts Using Biomorphic Explorers*. Jet Propulsion Laboratory, 4800 Oak Grove Drive, Pasadena, CA 91109. Lunar and Planetary Science XXX (1999).
9. Scott, G.P.; Ellery, A. *Biomimicry as Applied to Space Robotics with Specific Reference to the Martian Environment*. TAROS (2004).
10. Thakoor, S. 1st NASA/JPL Workshop on Biomorphic Explorers for Future Missions. NASA's Jet Propulsion Laboratory Auditorium. 4800 Oak Grove Drive, Pasadena, CA 91109, August 19–20, 1998.
11. Hirata, C. *A New Plan for Sending Humans to Mars: The Mars Society Mission*. California Institute of Technology (1999).
12. Clément, G.; Pavy-Le Traon, A. Centrifugation as a Countermeasure during Actual and Simulated Spaceflight: A Review. *Eur. J. Appl. Physiol.*, **92**, 235–248 (2004).

Section III

Preparing for Launch

Having described the challenges of training in the previous section, the objective of this final section is to follow the astronauts through their final 10 weeks of pre-launch preparation. Before describing the events leading up to launch, however, it is necessary to consider the new launch vehicle and spacecraft that will carry astronauts to orbit, and acknowledge the teams that work behind the scenes to make the mission a reality.

The new class of post-Shuttle astronauts will fly to the International Space Station (ISS), the Moon, and Mars onboard *Orion*, a spacecraft reminiscent of the Apollo capsule. Helping the new astronauts make their first journey into space are hundreds of people working to make the mission a success. In Chapter 9, readers are introduced to NASA's new family of launch vehicles and spacecraft before being introduced to some of the key personnel responsible for sending astronauts on their way. Chapter 10 picks up the pre-launch preparations with just 10 weeks remaining. Here, readers follow the astronauts as their schedule becomes ever more hectic with simulation following simulation, and the crew look forward to the impending launch with renewed anticipation. Finally, in Chapter 11, which begins with launch just days away and the astronauts in quarantine, the sequence of events comprising countdown is described – events that ultimately culminate in the realization of a lifelong dream.

9

The spacecraft and the launch teams

As NASA moves forward with its ambitious plan to maintain the ISS, revisit the Moon and, ultimately, send people to Mars, its blueprint for the future rests with a new family of launch vehicles and spacecraft. The new space hardware includes the *Orion* spacecraft and the *Ares I* rocket to launch it. Initially, *Ares I* will replace the Shuttle to ferry astronauts to the ISS. For Moon shots and longer missions, NASA's vision calls for a second launch vehicle to supplement the *Ares I*. Dubbed *Ares V*, this massive unmanned vehicle, due to be tested in 2018, will do the heavy lifting necessary to transport hardware to the Moon such as the lunar lander.

LAUNCH VEHICLES

Ares I

NASA's successor to the Space Shuttle is *Ares I* (Figure 9.1), an inline, two-stage rocket configuration topped by the *Orion* crew vehicle, its service module, and a launch abort system (LAS). In addition to its primary mission of ferrying four to six astronauts to the ISS, *Ares I* may also be used to deliver up to 25 tonnes of resources and supplies to astronauts in low Earth orbit (LEO) or park payloads in orbit for retrieval by spacecraft en route to the Moon or Mars.

First stage

The *Ares I* first stage is a single, five-segment reusable solid rocket booster (SRB), derived from the Space Shuttle program. Like the Shuttle's SRB, the Ares variant burns a specially formulated solid propellant called polybutadiene acrylonitrile (PBAN). On top of the upper segment, a forward adapter, called a frustrum (Figure 9.1), mates the vehicle's first stage with the second. To disconnect the stages during ascent, booster separation motors will fire. During launch, the first stage powers the

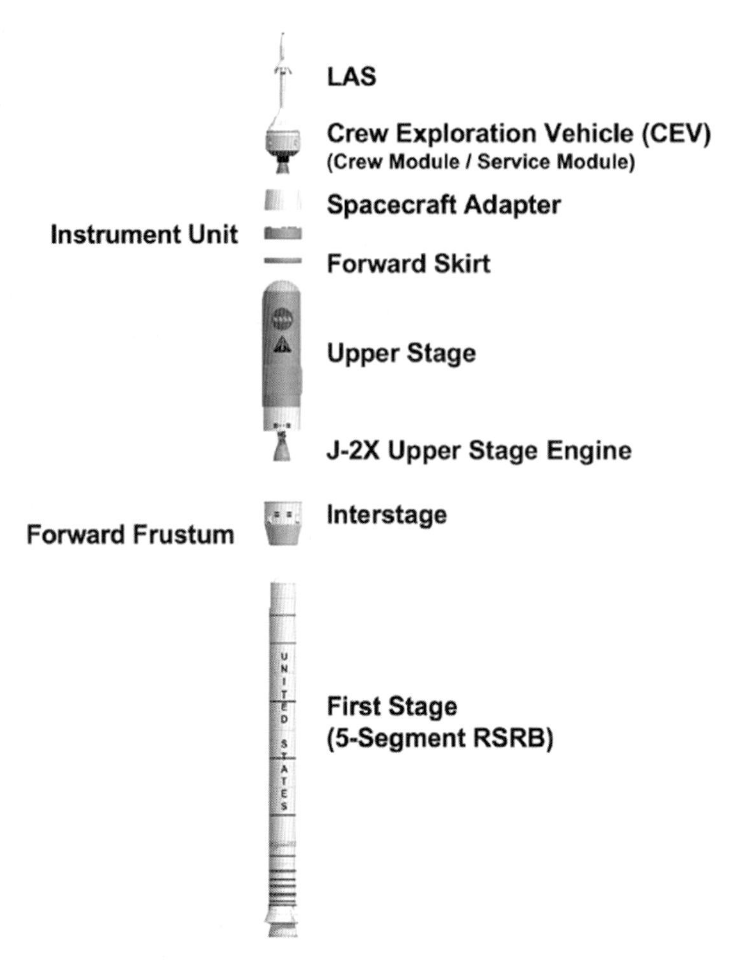

Figure 9.1. Exploded view of *Ares I*. Image courtesy: NASA.

vehicle toward LEO and, following a burn time of 126 sec, the reusable booster separates and the upper stage's J-2X (Figure 9.2) engine ignites, placing the vehicle into a circular orbit.

Upper stage

The *Ares I* upper stage (second stage) is propelled by the liquid oxygen (LOX) and liquid hydrogen (LH2)-fuelled J-2X engine. Approximately 133 sec after lift-off, the upper stage will separate from the first stage and the J-2X will ignite. The engine will burn for approximately 465 sec, burning more than 137,000 kg propellant, before shutting down at an altitude of 133.5 km. Shortly before the J-2X engine ceases operation, *Orion* will separate from the upper stage, whereupon *Orion*'s engine will

Figure 9.2. J-2X engine. Image courtesy: NASA.

ignite to insert the capsule into LEO. Following separation from *Orion*, the upper stage will re-enter the Earth's atmosphere and splash down in the Indian Ocean.

Ares V

The goals of NASA's Constellation Program include returning astronauts to the Moon, and onward to Mars. Realizing these missions requires the *Ares V*, NASA's new cargo launch vehicle, which will serve as the primary vehicle for delivering large-scale hardware to space such as lunar landing craft and materials for establishing a Moon base. The *Ares V* (Figure 9.3) is a two-stage, vertically stacked launch vehicle capable of carrying 188 tonnes to LEO and 71 tonnes to the Moon.

First stage and Core Stage

To achieve orbital insertion, the *Ares V* first stage relies on two five-and-a-half-segment SRBs derived from the Space Shuttle and similar to the SRB used to power *Ares I*. Flanking the two SRBs is a single liquid-fuelled central booster element, known as the Core Stage. Derived from the Shuttle, the Core Stage delivers LOX and LH2 to a cluster of six RS-68B rocket engines.

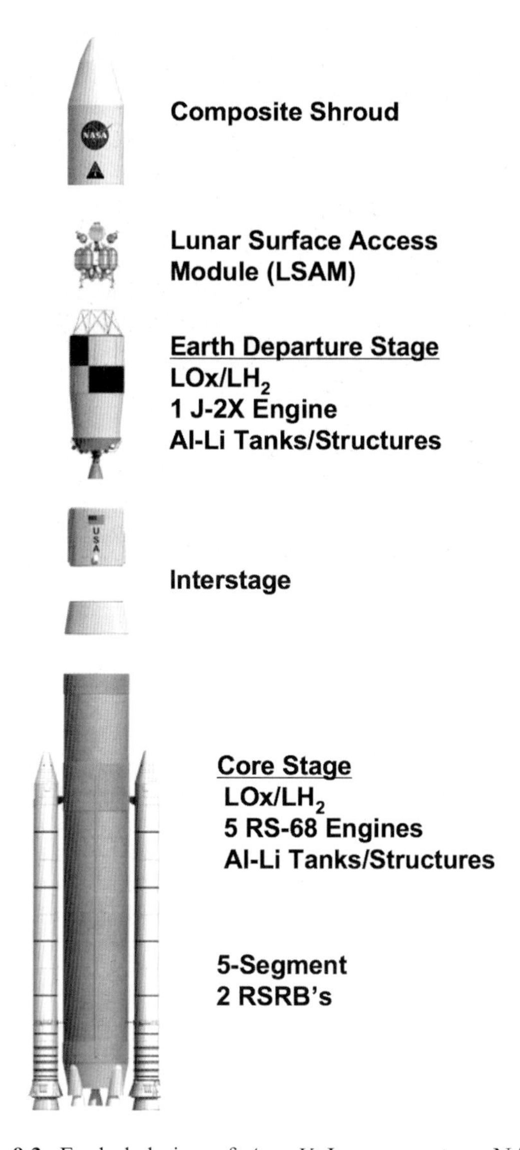

Figure 9.3. Exploded view of *Ares V*. Image courtesy: NASA.

Interstage and departure stage

Sitting on top of the Core Stage is an interstage, which includes booster separation motors. The interstage connects the Core Stage to the Earth Departure Stage (EDS), which is propelled by a LOX/LH2-powered J-2X engine. Anchored above the EDS is a shroud protecting the *Altair* lunar lander (Figure 9.4) that will transport astronauts to the lunar surface and the ascent stage that will return them to lunar orbit.

Figure 9.4. Lunar surface access module/*Altair*. Image courtesy: NASA.

Concept of operations

The SRBs and Core Stage power the *Ares V* to LEO. Following separation from the spent Core Stage, the EDS's J-2X engine provides the power to place the vehicle into a circular orbit, whereupon the departure stage shroud separates to prepare the lander for rendezvous with *Orion*. Shortly after the EDS has achieved orbit, *Ares I* delivers *Orion* to space and *Orion* docks with the departure stage and its lunar lander payload. Following docking, the EDS fires the J-2X engine a second time to achieve trans-lunar insertion (TLI), after which the EDS is jettisoned.

SPACECRAFT

Orion

The new spacecraft that will carry astronauts to the ISS, the Moon, and eventually Mars is called *Orion* (Figure 9.5). While it shares the conical shape of the Apollo capsule, it is one-and-a-half times as wide (5 m), and has more than double the habitable internal volume, enabling it to carry up to six astronauts to the ISS and four to the Moon. *Orion* also features a number of cutting-edge technologies, such as automatic docking and the ability to autonomously loiter in lunar orbit for up to 6 months. Thanks to its dual-fault tolerant avionics, based on those of the Boeing 787, *Orion* has the capability to sustain two computer failures and still return to Earth.

Figure 9.5. *Orion.* Image courtesy: NASA.

Additional safety is assured by the placement of *Orion* on top of *Ares I*. This means the capsule is not only protected from falling debris, but also permits a LAS that can blast the capsule to safety.

Launch abort system: launch

The LAS (Figure 9.6), which fits like a glove over the top of *Orion*, is designed to deal with the mission profile's most critical failure modes: lift-off and max-Q,* which occurs approximately 1 min after lift-off when the vehicle is traveling at Mach 2 (2,124 km/hr) at an altitude of 21 km.

The heart of the LAS is the solid-fuel rocket, with four outward-canted reverse-flow nozzles at its apex. In the event of an imminent fireball engulfing the pad, the rocket automatically fires for 2 sec, delivering a 15-G shock hauling *Orion* off the top

---

* Max Q is the point of maximum dynamic pressure or the point at which aerodynamic stress on a spacecraft in flight is at its peak.

Figure 9.6. Launch abort system. Image courtesy: NASA.

of *Ares I* at a speed of almost 1,000 km/hr and to an altitude of almost 2 km. Once clear of the danger area, eight attitude thrusters and two adjustable canards steer *Orion* east from the launch pad and parachutes deploy for a splashdown.

Launch abort system: Max-Q

Ensuring astronaut safety during the Max-Q phase of the mission profile is a little more challenging due to the aerodynamic drag and the shockwave-induced suction between the capsule and the service module behind it. However, engineers believe the system will be sufficiently robust to pull the capsule clear, after which *Orion* would right itself and descend before chute deployment at an altitude of 8,000 m.

Thermal protection system

Another important element of *Orion*'s design is its thermal protection system (TPS). Re-entering the atmosphere generates intense heat from air compression ahead of the spacecraft's supersonic shock wave. Whereas the Shuttle's re-entry speed was

27,800 km/hr (about Mach 23), *Orion*'s re-entry speed from lunar missions will be 38,600 km/hr (about Mach 31.5!), or about 40% faster than the Shuttle. At such high speeds, the heat buildup rate is five times greater than the Shuttle's, with temperatures reaching 2,650°C! For engineers working on *Orion*'s TPS Advanced Development Project, designing the heatshield was a huge challenge, but after evaluating eight candidate materials, an ablative system was chosen. Avcoat, which is made of silica fibers with an epoxy-novalic resin filled in a fiberglass-phenolic honeycomb, is manufactured directly onto *Orion*'s TPS substructure and attached as a unit to the capsule during assembly. The use of an ablative system, in which parts of the TPS burn away (or ablate), mirrors that of the Apollo Program's approach, in which the entire entry capsule was covered with an ablator.

Reusability

The Shuttle was originally designed for 100 flights each, and NASA had envisioned a flight rate of 60 missions a year, but it didn't happen. The Shuttle never flew more than nine missions in a calendar year, which meant economies of scale were never realized. Although *Orion* will be reusable, it will be less reusable than the Shuttle. While the main capsule is designed to be reused (five to ten times), the service module, heatshield, and the LAS must be replaced after every mission.

Altair

Altair will be capable of landing four astronauts on the Moon and provide life support and a base for week-long initial surface exploration missions, before returning the crew to the *Orion* spacecraft that will bring them home to Earth. *Altair* will launch aboard an *Ares V* rocket into LEO, where it will rendezvous with *Orion*. The three primary elements of the lunar lander include a LOX/LH2-powered descent stage, a hypergolic-powered ascent stage, and a LOX/LH2-powered descent module that will provide propulsion for powered descent to the surface and serve as a platform for lunar landing and lift-off of the ascent module.

MISSION PROFILE

During the first 2.5 min of flight, the SRB will power the vehicle to an altitude of about 57 km at a speed of Mach 5.7. After its propellant is spent, the SRB will separate and the upper stage's J-2X engine will ignite and power *Orion* to an altitude of about 130 km. Then, the upper stage will separate and *Orion*'s service module propulsion system will complete the trip to a circular orbit of 297 km. Once in LEO, *Orion* and its service module will rendezvous and dock either with the ISS or with the *Altair* lunar lander and Earth departure stage that will carry the crew to the Moon. Once they have reached lunar orbit, astronauts will use *Altair* to travel to the lunar surface while *Orion* remains in lunar orbit for up to 210 days, awaiting return of the crew.

Following their surface stay, the crew will use *Altair*'s ascent vehicle to return to lunar orbit and reunite with *Orion*. The service module main engine will then provide the power for the trans-Earth insertion (TEI) burn, enabling *Orion* to break out of lunar orbit and return to Earth. The service module will support the crew module until the two modules separate just before re-entering Earth's atmosphere. *Orion* will re-enter Earth's atmosphere and, with the use of parachutes, return the crew to Earth.

The *Orion*/*Ares I*/*Ares V* configuration has been labeled by former NASA Administrator, Michael Griffin, as Apollo on Steroids. Although *Orion* and *Ares I* feature improvements over the Apollo capsule and the *Saturn V* rocket, many are disappointed the return to the Moon will not feature more new technology, but, given NASA's cost and scheduling constraints, the *Orion*/*Ares I* configuration is the best option. Of more concern is the impending 5-year hiatus (which may be reduced if commercial operators such as SpaceX can man-rate their vehicles – Figure 9.7) between the retirement of the Shuttle and the operational status of *Orion*/*Ares I*, a gap requiring the new class of astronauts to hitch a ride with the Russians onboard their Soyuz.

Figure 9.7. SpaceX's Dragon capsule may hold the key to reducing the hiatus in American manned spaceflight capability following retirement of the Space Shuttle in 2010. Image courtesy: SpaceX.

Figure 9.8. Following the Shuttle's retirement, American astronauts will hitch rides onboard the Russian Soyuz capsule at $51 million a ticket! Image courtesy: NASA.

Soyuz

A Soyuz capsule (Figure 9.8) first ferried a crew to the ISS in November, 2000. Since that time, at least one Soyuz has been berthed at the ISS, serving as a lifeboat in the event the crew have to return to Earth unexpectedly. Following the *Columbia* accident in February, 2003, the Soyuz also became the means of transportation for crewmembers going to or returning from the ISS.

Orbital module

This element of the spacecraft is used by three crewmembers while on orbit during free-flight. It features a docking mechanism with hatch and rendezvous antennas located at the front end. The docking mechanism is used to dock with the ISS and the hatch allows entry into the station. The rendezvous antennas are used by the automated docking system to maneuver towards the station for docking.

Instrumentation/Propulsion module

This module comprises an Intermediate, Instrumentation, and Propulsion compartment. The Intermediate Compartment connects to the Descent Module, while the

Instrumentation Compartment contains oxygen storage tanks, attitude control thrusters, and communications and control equipment. Also housed within the Instrumentation Compartment are the Soyuz's primary guidance, navigation, control, and computer systems. The Propulsion Compartment contains the primary thermal control system and the radiator. The propulsion system, batteries, solar arrays, radiator, and structural connection to the Soyuz launch rocket are also located in this compartment. Following the final de-orbit maneuver, the Instrumentation/Propulsion module separates from the Descent Module and burns up in the atmosphere.

Descent Module

The opposite end of the Orbital Module connects to the Descent Module via a pressurized hatch. Following the de-orbit maneuver, the Orbital Module separates from the Descent Module and burns up in the atmosphere. Since the Descent Module is where the cosmonauts/astronauts sit for launch, re-entry, and landing, the capsule's controls and displays are located here. The module also contains the primary and backup parachutes and landing rockets.

Concept of operations

The Soyuz is launched from the Baikonur Cosmodrome in Kazakhstan aboard a Soyuz rocket. Once the Soyuz reaches orbit, it spends 2 days "chasing" the ISS. Prior to the final rendezvous phase, crewmembers don pressurized suits and monitor the automated docking sequence. While the rendezvous and docking are automated, the crew has the capability to manually execute these operations. Once docking is complete, the crew equalize the air pressure of the Soyuz with the ISS before opening the hatches.

LAUNCH TEAMS

The nerve-center during launch is NASA's Launch Control Center (LCC), a four-story building located at Kennedy Space Center (KSC) used for the supervision of launches from Launch Complex 39. Attached to the southeast corner of the Vehicle Assembly Building (VAB), the LCC contains telemetry, tracking and instrumentation equipment, the automated Launch Processing System (LPS), and four firing rooms. It is inside the firing rooms (Panel 9.1) that the launch team (Table 9.1) make the key decisions that will launch the rookie astronauts on their first spaceflight. Following "call to stations", the launch team report to their respective computer/communication consoles for the final countdown and share the responsibility and excitement of launching a manned space mission.

Panel 9.1. Firing rooms

In preparation toward supporting future launches for NASA's Constellation Program, the agency is reconfiguring some LCC facilities. For example, the LCC is being redesigned to use only 50 launch controllers compared with the nearly 200 required for a Space Shuttle launch. Also, Firing Room 1 has already been transferred to the Constellation Program for reuse and is being rebuilt to support systems required for the Ares rockets and *Orion* spacecraft. Firing Room 1 already holds the distinction of overseeing the maiden launch of the Space Shuttle, and was recently renamed the Young/Crippen Firing Room in honor of Commander John Young and pilot, Robert Crippen, as a tribute to their first Shuttle flight.

Table 9.1. Primary firing room positions.

Position	*Acronym*	*Description*
Launch Director	LD	Head of launch team. Responsible for making final "go/no go" launch decision after polling team members
Flow Director	FD	Responsible for preparation of vehicle for launch, and remains in LCC in an advisory capacity
NASA Test Director	NTD	Responsible for all pre-launch testing, whether involving the flight crew, the vehicle, or ground support equipment. Also responsible for safety of personnel on the pad after fuelling. Reports to LD
Test Conductor	TC	In charge of pre-flight checkout and testing of vehicle. Manages engineers in the firing room who monitor the vehicle's systems
Launch Processing System Coordinator	LPS	Monitors and controls most vehicle assembly, checkout, and launch operations
Support Test Manager	STM	Manages and integrates ground support resources in support of launch countdown
Safety Console Coordinator	SAFETY	Responsible for ensuring all ground safety requirements affecting personnel and flight hardware safety are met during launch countdown activities through verification of appropriate safety measures
Landing and Recovery Director	LRD	Manages KSC assets to support landing operations and SRB recovery. Provides coordination for any contingency landing operations with Department of Defense Manned Space Flight Support (DDMS) office and the JSC Landing Support Officer (LSO)
Superintendent of Range Operations	SRO	Ensures downrange airspace and splashdown areas remain clear for launch, and monitors weather near the launch site

Table 9.1. *continued*

Position	*Acronym*	*Description*
Ground Launch Sequencer Engineer	GLSE	Responsible for monitoring operation of the automated Ground Launch Sequencer system, which controls the countdown from T–9 min until launch. After T–9 min. through to T–31 sec, they are in charge of implementing a manual hold if necessary

Launch Director

Leading the team is the Launch Director. Many people think the Launch Director's job is to get the launch vehicle off the ground, but it isn't. The Launch Director's job is not to launch the launch vehicle, but to say "no" if something doesn't look or feel exactly right, even when the rest of the launch team say "go". Ultimately, it is the Launch Director's job to make sure the launch vehicle stays on the ground in the event of what NASA euphemistically refers to as an "off-nominal event". Sometimes, weather may prompt the Launch Director to wave off the launch; sometimes it may be a minor technical glitch, but, no matter what the problem, the Launch Director has the last say. Given the magnitude of such a decision, it isn't surprising the leaders of the launch team are calm, assured individuals who possess a natural reserve that keeps them from becoming excited even when others might be heading for a coronary! The pressure of responsibility is compounded by the enormity of the stage, since the Launch Director is the focal point during the countdown, his words being transmitted live to millions around the world.

To ensure the countdown proceeds smoothly on launch day, the Launch Director and his team of test directors, controllers, and engineers will go through at least two simulated countdowns before launch day. During the simulated countdowns, 20–30 simulated problems are thrown into the mix, with the intent of letting the team make mistakes. The training philosophy of these simulated countdowns is that the best way to learn something is to make a mistake, correct it, and move on. The philosophy has been tried and tested over more than 100 Space Shuttle launches and is part of the reason the pressure cooker environment of the LCC is an oasis of calm come launch day.

In addition to supervising his team, the Launch Director must have a basic knowledge of all the launch vehicle's systems, an understanding of orbital mechanics, *and* be capable of factoring in weather and abort requirements. In addition to his/her launch duties, the Launch Director also serves as chairman of the Rapid Response Team (RRT), a duty requiring him/her to coordinate the network of NASA first responders in the event of a dreaded off-nominal event during lift-off or landing. The job is so unfathomably complex that few people outside of the LCC really have an idea of what the Launch Director does. Despite the pressures, for the Launch

Director, entrusted with launching half a dozen crewmembers on a rocket with millions of pounds of thrust, the job is the best in the world, although astronauts might think otherwise!

Mission Control

While the LCC is responsible for the events leading up to launch, the Flight Director hands over control to the Mission Control Center (MCC) at Johnson Space Center (JSC) once the vehicle has cleared the tower (at about T + 7 sec). Since the final chapter of this book follows the astronauts until they are in LEO, it is useful to understand the MCC's role.

The MCC, a windowless three-story building (Building 30), has two Flight Control Rooms (FCRs – usually pronounced "fickers"), from which manned space flights are managed. In the FCRs, teams of flight controllers work 8-hr shifts, directing and monitoring all aspects of flight. Some of the most prominent command and control positions are listed in Table 9.2.

Table 9.2. Key Mission Control positions.

Position	*Acronym*	*Description*
Flight Director	FD	Leader of the flight control team. Responsible for mission operations and decisions relating to safety and flight conduct
Spacecraft Communicator	CapCom	Primary communicator between MCC and spacecraft
Flight Dynamics Officer	FDO	Plans vehicle's maneuvers and follows vehicle's flight trajectory along with the Guidance Officer
Guidance Officer	GDO	Responsible for monitoring spacecraft's navigation and guidance computer software
Flight Surgeon	Surgeon	Monitors crew activities and provides medical con sultations with crew as required. Also keeps Flight Director informed on the state of crew health

The LCC and MCC personnel represent the leading edge of a huge network of behind-the-scenes support that extends far beyond Houston. Without this complex network, no mission could succeed – a fact no one is more aware of than the astronauts as they begin their final preparations for launch.

10

Ten weeks and counting

Like an elite athlete tapering for a race, the astronauts, with just a few weeks to go before launch, are approaching the peak of their training. As the finish line approaches, their preparation becomes ever more intense as their lives become a blur of simulators and final medical checks. In addition to confirming their proficiency, the simulators get the crew into the "mode of flying mindset" and focused ever more on the final test – the ride into space!

PRACTICE MAKES PERFECT

As the launch date approaches, crews spend an increasing proportion of their time in fixed-base and motion-based simulators in the Space Vehicle Mock-up Facility (SVMF), conducting dress rehearsals for the mission. The fixed-base simulator, which replicates *Orion*'s flight deck, is usually used to teach orbit operations. Astronauts seated in the fixed-base simulator are presented with convincing views of Earth and space and, for added realism, the systems function just as they would in the real spacecraft (they even make the same noise as they would in orbit!). In contrast, the motion-based simulator is used to practice ascent and re-entry skills and is housed on hydraulic jacks to give the crew an accurate simulation of the launch and landing. At the beginning of simulator training, astronauts are required only to perform stand-alone sessions. The stand-alone sessions, which typically last 5 or 6 hr, help the crew rehearse a portion of the mission on their own under the guidance of instructors. Once an astronaut has acquired a thorough knowledge of a particular phase of the mission, the instructors "help" by deliberately interfering during the simulation. These premeditated acts of sabotage include malfunctions, such as a failed engine prior to de-orbit burn or cabin depressurization. As the crew progresses, the stand-alone simulations become progressively more difficult as crewmembers begin to interact not only with their assigned Simulation Supervisor (SimSup in NASAese), but also with flight controllers and other astronauts. Once the crew has completed the stand-alone simulation phase, they move on to the integrated simulations (known in NASA parlance as *integrated sims*), which is where this chapter begins. But, before describing the integrated sim phase, it is necessary to introduce the instructors tasked with fine-tuning the skills of those preparing for space.

CHOREOGRAPHING CHAOS

In Spaceflight Training and Facility Operations (STFO), instructors (Panel 10.1) train the crew and the flight controllers. Here, integrated sims involving the flight crew and the flight controllers are orchestrated by a highly qualified cadre of instructors. Their job is to make life miserable by presenting all manner of intricate failure scenarios and ensuring a steady stream of system failures! To achieve this, the STFO team work closely with the crew in the cockpit simulator and those stationed in the Mission Control Center (MCC). While seated in their simulator, primary flight displays (PFDs) feed the astronauts computer-generated images of what would appear during an actual mission, while flight controllers see the data on their MCC workstations, just as they would do during flight. By training flight controllers and crew to react to the data as they would during the actual mission, the training focuses both systems expertise and the communication and coordination between those working on the ground and the crew in space.

Panel 10.1. Simulation instructors' responsibilities

The simulation team is headed by the Team Lead. He/she leads a team of four to seven instructors who work in a room near the Mission Simulator. The Mission Simulator team monitors the crew's actions and introduces malfunctions during a given scenario.

The SimSup leads a team of four to seven instructors who work in the Simulation Control Area (SCA) within the MCC. The SCA team generates scenarios, works with the Mission Simulator team ensuring scenarios are introduced correctly, and monitors the flight control team's decisions and actions.

For simulations involving the Space Station, the Station Training Lead leads a team of instructors who work in an Instructor Station near the Space Station Training Facility (SSTF). The SSTF team generates the scenarios and monitors both the crew and Space Station flight controller actions and introduces malfunctions during a given scenario.

The Team Lead, SimSup, and Station Training Lead communicate over voice loops during the simulation, ensuring everyone is aware of the situations and keeps the teams focused on the overall objectives.

Prior to conducting an integrated sim, the SimSup and the SCA team sit down and generate scenarios for a day's activity, which is a little bit like writing a script. After the scripts have been written, the SimSup combines them into a script package that will be used for a specific sim. The script package is then distributed to all teams involved in the sim, such as the MCC. In addition to integrated sims being more comprehensive than stand-alone sims, there are other notable differences. For example, during stand-alone sims, when the MCC is not involved, the vehicle mission simulator team generates the scenarios and executes them while pretending to be MCC. In contrast, during an integrated sim, the SCA team generates scenarios, with the vehicle simulator team implementing the scenarios' malfunctions, while the SCA team observes and evaluates the flight control team's response.

Leading the simulation is the SimSup, who must interface with not only the flight control team(s), but also all of the communications and the network systems. The SimSup is also the person who can throw in last-minute glitches, such as breaking a certain communications link or causing a radar malfunction. However, while it might sound like fun to try and trip up astronauts, the simulation instructors script their scenarios very carefully, ensuring everyone is thinking on a larger scale. For example, failures are not introduced to the timeline randomly, but are initiated with the aim of stimulating team problem-solving skills. While some failures may not have any consequence for any system, the instructors often throw in failures that affect many systems at once, which really messes up the timeline and demands close coordination of the crew. While it may appear simulation instructors get paid to be devious, their objective is really just to test the crew's skill and knowledge of spacecraft systems in creative ways they don't anticipate. By exercising their malfunction creativity, simulation instructors not only ensure the crew focus on the skills and knowledge necessary for the mission, but also train the crew to correct and/or recognize off-nominal conditions in onboard systems. To achieve this objective, simulation instructors must throw ever trickier curve balls, which is why most sims are more intense than an actual mission. The method behind the madness is clear: while everyone's ultimate goal is a flawless mission, the best way to get there is not through a flawless simulation.

MARATHON-INTEGRATED SIM

The marathon sim (which sometimes last 36 hr!) usually takes place some time during the last 10 weeks preceding launch. For the crew, the marathon sim is as much a test of endurance as skill and proficiency, as the final weeks leading to launch are some of the most grueling of all. With work weeks routinely running to 60 hr or more, the astronaut's training schedule is a blur of sims, EVA preparation, procedure training, and briefings. Quality time with the family becomes a distant memory, routine desk work is forgotten, and lunch-breaks are reduced from half an hour to 10 min (or less!) as the mission quickly takes over the crewmember's life.

During the marathon sim, the crew divides its time between the vehicle mission simulator and the Space Station Training Facility (SSTF). In addition to the simulators occupied by the astronauts, the Space Station Flight Control Room, Simulation Control Room, and several "back rooms" are also integrated into the dress rehearsal. While the astronauts spend every hour of the sim in the simulators, the flight controllers and simulation instructors rotate in teams around the clock. The first 24 hr reviews the first 2 days of the mission, after which ISS flight controllers join the sim for rendezvous and docking activities. Obviously, planning a sim lasting 36 hr takes weeks of careful preparation. In common with the stand-alone sims, the marathon-integrated sim is full of selected errors, malfunctions, and mishaps, each designed to test the crew and the flight controllers. However, the script serves only as a guideline and can be changed with a few keystrokes or a few clicks of a mouse. As the instructors throw wrenches into the sim, the crew work their way through procedures designed to solve the glitches, until, finally, the sim comes to an end. As soon as the sim is over, the astronauts review the lessons learned and jump back into their frenetic schedule.

FLIGHT SURGEON PHYSICAL EXAM

Thirty days before launch, astronauts undergo a medical examination, usually referred to as the L-30 exam. The examination (Table 10.1) is a routine physical taking no more than 60 min. As with all medical examinations astronauts must undergo, the data from the L-30 exam are entered into NASA's Longitudinal Study of Astronaut Health (LSAH), an ongoing research protocol that began in 1992 to examine the long-term physiological effects of space flight on astronauts.

Table 10.1. L-30 flight surgeon physical examination for long-duration crews.

Test	*Description*
Vital signs	• Pulse and blood pressure recumbent, sitting, standing • Body temperature • Respiratory rate • Height and weight
Head and face	• Nasal mucosa • Sinuses, maxillary, and frontal
Mouth and throat	• General examination
Ears	• External meatus, tympanic membrane • Response to Valsalva
Eyes	• General appearance • Extra-ocular movements • Pupil reactivity • Ophthalmoscopic exam

Table 10.1. *continued*

Test	*Description*
Neck	• Thyroid • Vascular exam • Motion
Chest and lungs	• Cardiovascular exam: ○ Cardiac auscultations ○ Carotid and venous upstrokes ○ Peripheral pulses
Abdomen	• Auscultation • Palpitation of major organs and herniations
Rectum and anus	• Prostate exam for males • Rectal vault • Occult blood testing
Genitourinary	• Appearance • General exam • Herniations
Breast exam	
Pelvic exam	• Pap smear for female crewmember
Extremities	• Range-of-motion • General strength assessments of 1–5 scale
Spine	• General appearance and mobility
Skin	• Lymphatics • Identifying body marks
Neurologic	• Standard functional and gait exam

PACKING FOR SPACE

Packing for a 6-month ISS mission is little different from packing for a camping trip or a holiday abroad. Astronauts require changes of clothing, toothpaste, shampoo, deodorant, music, and even i-Pods! Helping the astronauts pack is a Flight Equipment Processing Associate (FEPA), a United Space Alliance (USA) employee, whose job it is to pack all the crew's clothing, crew preferences, hygiene and personal items in a meticulous and space-saving fashion. For astronauts who regularly log 60-hr work weeks as the launch date approaches, the FEPA is a godsend. While packing for an ISS increment might not appear to be a particularly time-consuming chore, anything that flies into space is subject to a labyrinthine series of flight-processing checks and requirements. To assist the crew in deciding which items to take, there is a Crew Options List (COL). Some items on the COL include watches, handkerchiefs, seat cushions, and safety helmets. These items are usually placed in the astronaut's personal belongings locker onboard the spacecraft. Astronauts have been known to bring along pennants with their college name on them, shirts, favorite books, i-Pods loaded with their favorite music (Table 10.2), and special items from family members and friends.

Crew members also have a Personal Preference Kit (PPK) in which they can fly up to 20 personal items, although these items must be very small because the bag in which they are carried isn't much larger than a tennis ball! For larger items, each mission flies the Official Flight Kit (OFK), comprising a list of items approved through official NASA channels. Items in the OFK (Table 10.3) are usually flown on requests from foreign governments and professional organizations. Some of the larger items flown in the OFK have included flags of foreign countries, patches, or other special awards that are presented later to honorees of an organization. To make sure the crew's clothing and item preferences are handled correctly, the FEPA works extensively with crew representatives and US engineers. Once all the items have been selected and passed muster, the FEPA arranges for a bench review a month prior to launch, which allows the crew a final opportunity to see how their clothing and personal items are packed for flight and perhaps to make any last-minute changes.

Table 10.2. Robert Thirsk's music selection (ISS Expedition 20/21, 2009).

Artist	*Album*
Jackson Browne	Running on Empty
Rolling Stones	The Best of the Rolling Stones
Van Morrison	The Best of Van Morrison
The Alan Parsons Project	Collection
Radiohead	OK Computer
The Beatles	Abbey Road
Diana Krall	Live in Paris
Jennifer Warnes	Famous Blue Raincoat
Joni Mitchell	Blue
Feist	Let it Die
Willie Nelson	Stardust
Vivaldi	The Four Seasons
Holst	The Planets

Table 10.3. Robert Thirsk's official flight kit (ISS Expedition 20/21, 2009).

Organization represented	*Item description*
National Research Council	3-D aluminum replica of the NRC badge
Montreal Neurological Institute	No. 4 Penfield dissector. The dissector is a commonly used neurosurgical instrument. A physician himself, Bob took the dissector into space as a symbol of medical innovation
Prime Minister's Award	Teaching Excellence Award Pin. Taken in honor of the educational value of his mission
Canadian Space Agency	Interferometer Instrument. The instrument was first flown in 1984 on Marc Garneau's (for whom Bob was backup) mission
Canadian Universities	Bob performed several research experiments onboard the ISS and brought a patch for each experiment he conducted
Canada	Canadian Flag, which remained on ISS
Glenbow Museum	Artifact – cap badge of Captain May, a Canadian World War I flying ace and pioneer bush pilot
Manitoba Museum	Arctic Discoveries Medal
University of Calgary	Iron ring, in honor of Dan Mercier, a former classmate of Bob's
Canada Council for the Arts	Two Governor General Literary Award-winning books as a symbol of the value of literacy

TERMINAL COUNTDOWN DEMONSTRATION TEST

A couple of weeks before launch, the crew fly from JSC to KSC for the Terminal Countdown Demonstration Test (TCDT), the final dress rehearsal for launch that follows almost all the steps of an actual launch except for tank fuelling and the actual ignition sequence. Before the TCDT begins, however, the crew has a busy schedule of other activities, one of which is learning how to drive an armored personnel carrier (APC)!

Driving the M-113 escape tank

The M-113 (Figure 10.1) is used by the pad rescue team to extract flight crew members, closeout members, or final instruction team members during a launch flow. First developed in the 1960s, NASA has four of them, all of which are utilized on launch day. To become proficient in driving the M-113 takes about 15 min and each astronaut must be cleared to drive the vehicle in the event of what NASA likes to refer to as a Mode 1 Egress. A Mode 1 Egress is a self-egress in which the crew would exit the spacecraft and use a rollercoaster escape system to a bunker, before

Figure 10.1. Astronauts put NASA's M-113 escape tank through its paces before the Terminal Countdown Demonstration Test. Image courtesy: NASA.

driving an M-113 to safety. If all goes well, astronauts should be able to ingress the M-113 in less than 5 min.

After fulfilling many a kid's dream of driving a tank at high speed, the astronauts attend a fire and safety reminder briefing, reviewing the details of the different hazards in an emergency. For example, they are reminded which fuels they might be exposed to on the launch pad. Following the fire and safety briefing, they attend a security briefing, after which they meet up with the FEPA to check the items they requested during launch. The day often ends with dinner at the Beach House with some of the KSC Management team. The Beach House, a cozy house perched above the dunes at the edge of a pristine beach, is a building steeped in NASA history, and one that astronauts often visit with their families before launch.

Emergency egress procedures

The next day begins with lectures, mostly concerning the impending TCDT and the condition of the vehicle and the spacecraft. The lectures usually last until lunchtime. As with all meals, the astronauts eat in their crew quarters before being bussed to the launch pad for a half-hour press conference in front of the vehicle. Following the question-and-answer session, the crew rides the elevator to the launch tower and reviews the ingress and emergency egress procedures they will use during the TCDT the following day. One of the items of equipment they pay particular attention to is the emergency escape system (Figure 10.2 and Panel 10.2), which resembles a rollercoaster in more ways than one!

Figure 10.2. NASA's new emergency escape system. Note the rollercoaster system in the right of the image. Image courtesy: NASA.

Panel 10.2. NASA's rollercoaster emergency escape system

Following the retirement of the Space Shuttle in 2010, NASA will replace its current slide wire escape system with the Ares I Emergency Escape System (EES). The EES comprises three multi-passenger cars on a set of rails reminiscent of a rollercoaster.* Its purpose is to transport astronauts and ground crew quickly from the vehicle on the launch pad to a protective concrete bunker in case of an emergency. The railcars, each capable of holding six astronauts, would ride unpowered down the launch tower vertically (!) and be decelerated using a passive and magnetic friction braking system. When operational, the EES will be capable of transporting up to 12 personnel (six flight crewmembers and six closeout crew) from the loading platform to inside the bunker in less than 2 min with no assistance from fire/rescue personnel.

In theory, the EES will be able to transport astronauts to safety in less than 2 min. While this amount of time might be sufficient for a fuel leak scenario, for most emergency scenarios involving a fully fuelled rocket, the rollercoaster design just gives the astronauts something to think about before they die.

Terminal countdown demonstration test

The following day, the crew eats breakfast at 6.30 and goes directly to the Operations and Checkout Building, where they dress up in their pumpkin-colored spacesuits. At about 7.45, they assemble as a group and exit the building to the Astrovan. Even though the TCDT is only a practice, there are usually several dozen people waving and cheering. The drive to the pad takes between 10 and 15 min and is conducted in a convoy with flashing blue lights, just like it will be on launch day. Once in the "white room", the crew ingress the spacecraft, performing the same procedures they will perform in 3 weeks' time, when the launch will be for real. After lying on their backs for a couple of hours conducting communication tests and ticking off the countdown checks, the TCDT finishes with the crew performing a Mode 1 egress. In pairs, the astronauts exit the spacecraft and walk through the white room, past the elevators, and strap themselves into the rollercoaster seats. If the emergency was real, the astronauts would slide down the gantry at some speed and end up inside the bunker, where they would stay until the threat had passed or they would drive the M-113 to a pre-arranged helicopter pickup point. For the TCDT, it is considered too risky to practice!

---

* Apparently, one of the lead mechanical engineers at KSC is a rollercoaster fanatic, and some wonder if he had anything to do with the design choice!

After the TCDT, the astronauts head back to Houston. With less than 3 weeks to go, most of the training is over. For those conducting EVAs, there will be one final practice session in the Neutral Buoyancy Laboratory (NBL), and, for the pilots, there will be some fine-tuning of their skills in the simulator. Ten days before launch, astronauts undergo another medical examination (usually called the "L-10"), which takes about 30 min and assesses vital signs, ear, nose, and throat, chest and lungs, abdominal, extremities, and neurologic function. The L-10 exam also includes swabs and other tests to make sure crewmembers aren't infected with any illness. After the exam, NASA limits astronaut contact with other people, although formal quarantine starts 3 days later.

QUARANTINE

A week before launch, the astronauts enter quarantine in NASA's Astronaut Quarantine Facility (AQF), located at JSC. This precaution is taken before every mission, since the consequences of an astronaut getting sick are serious; even a common cold can have a mission impact if a crewmember is unable to clear their ears due to congestion, especially with the changes in pressure required for a spacewalk. Quarantine is strict, with the crew surgeon being isolated with the astronauts and anyone exhibiting signs of illness being prohibited from working with the crew.

The AQF (Figure 10.3) consists of 12 bedrooms with private baths, a large conference room, laundry facilities, a kitchen and dining facility, computer workstations, a workout room, a medical exam room for pre and postflight evaluations, and a lounge area for meetings with medically approved visitors. Although the pace of training slows down in quarantine, mission trainers still find time to schedule a launch simulation or two. The instructors keep a final series of launch and ascent runs for the very end to ensure the astronauts' reflexes are fine-tuned and ready for the real event. Usually, the final sim consists of four different launch scenarios and, as usual, the crew has to respond to all sorts of simulated

Figure 10.3. NASA's Astronaut Quarantine Facility. Image courtesy: NASA.

failures, but only during the first three. The fourth simulation is a completely normal ascent, which serves as a gift from the instructors to the crew. It's their way of wishing the crew good luck. For the crew, who are by now conditioned to expect a never-ending series of malfunctions, the normal ascent is a big surprise; after spending so much time in the sims training to respond to one glitch after another, most of the crew can't remember the last time they experienced a simulated launch without an alarm going off!

The astronauts stay in the AQF for 2 or 3 days before flying to the Cape, where they are subjected to the same restrictions in their Crew Quarters. Thanks to the reduction in the hectic pace of training, the astronauts finally have the opportunity to catch their collective breath. They take advantage of the precious time to catch up on paperwork, finalize their preflight preparations, review flight plans and procedures, and focus on the impending final count.

11

The final count

"Oh! I have slipped the surly bonds of Earth
And danced the skies on laughter-silvered wings;
Sunward I've climbed, and joined the tumbling mirth
Of sun-split clouds – and done a hundred things
You have not dreamed of – wheeled and soared and swung
High in the sunlit silence. Hovering there,

I've chased the shouting wind along, and flung
My eager craft through footless halls of air.
Up, up the long delirious burning blue
I've topped the wind-swept heights with easy grace
Where never lark, or even eagle flew.

And while with silent lifting mind I've trod
The high untrespassed sanctity of space,
Put out my hand and touched the face of God."

High Flight, John Gillespie Magee, Jr

LAUNCH DAY

For the astronauts, launch day begins with breakfast prepared by NASA dieticians. While some may have an image of astronauts enjoying a breakfast of steak, eggs, and coffee, the reality is somewhat different. First, most astronauts are too damn nervous to eat anything more than half a slice of toast and, second, due to the effects of having to lie down for 3 hr in the supine position, none of them so much as looks at a cup of coffee. In fact, many of the astronauts, embarrassed at the thought of having to use the Maximum Absorbency Garment (MAG), intentionally dehydrate themselves prior to launch. However, since the breakfast is also a mandatory photo opportunity for the press and the NASA media people, the astronauts usually fake a carefree smile and pretend to eat.

After the meal, the crew performs one final prelaunch briefing wearing their

Figure 11.1. Expedition 19 commander, cosmonaut Gennady Padalka (second left), poses with astronauts, Michael Barratt and Tim Kopra, during a preflight press conference at NASA's Johnson Space Center. Public Affairs Office moderator, Kylie Clem, is at left. The crew wear their traditional mission golf shirts during their prelaunch briefing. Image courtesy: NASA.

traditional mission golf shirts (Figure 11.1) before attending a teleconference to review the launch countdown status and weather forecast. The weather forecast isn't just for KSC, but also for the abort sites. For example, the vehicle can't be launched if the weather at the primary transatlantic abort site (this may be Istres in France) or alternative abort sites (such as Zaragoza in Spain) isn't cooperating. Next, the crew visit NASA flight surgeons for one final medical examination before visiting the bathroom for a final attempt to void their bladders before suiting up.

Suit-up

On launch day, the astronaut's first challenge isn't dealing with the acceleration stress following launch or adjusting to weightlessness, but donning the rather cumbersome launch and entry suit (LES). In fact, because the suit-up process is so complicated, the carefully choreographed countdown actually allows a 45-min window in case any problem occurs.

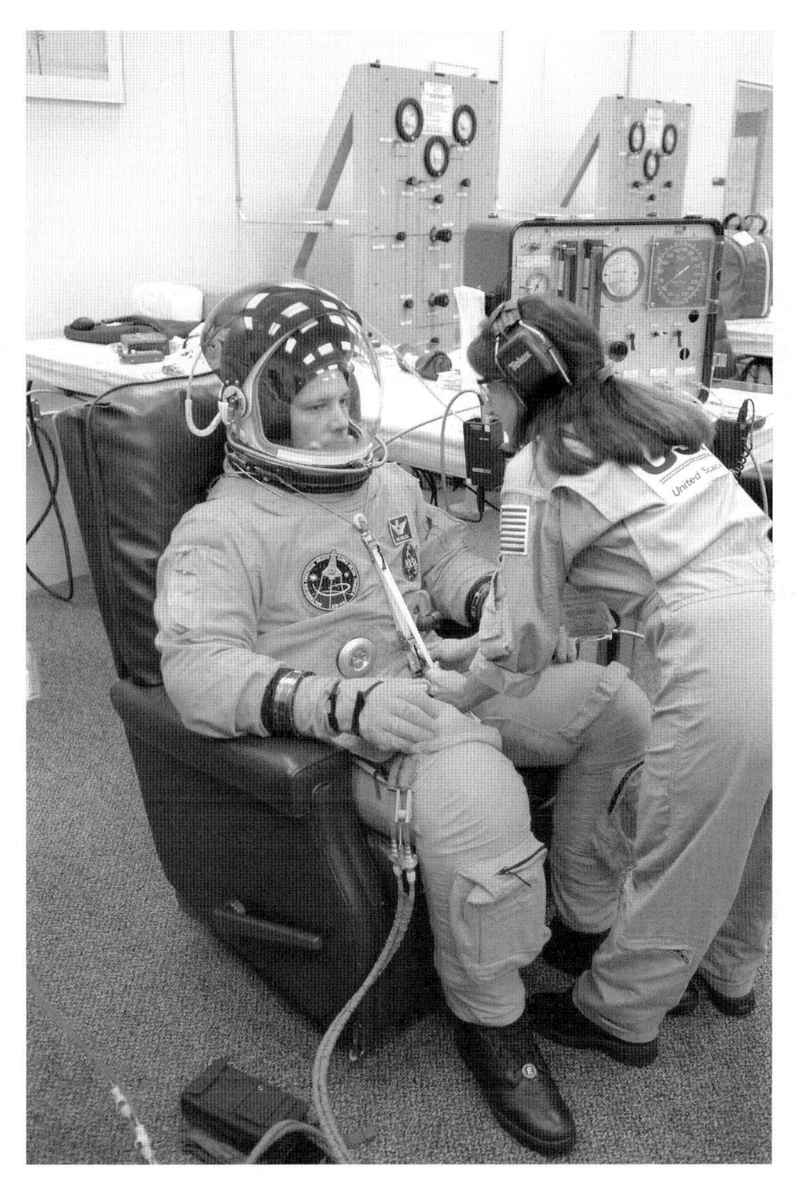

Figure 11.2. Astronauts suit up, assisted by a United Space Alliance suit technician. Image courtesy: NASA (*see colour section*).

Because the suits are so awkward to put on, each astronaut has their own personal suit technician who helps the soon-to-be spacefarer don their suit (Figure 11.2 and Panel 11.1). The process of suiting up starts early on launch day, when astronauts go to the suit-up room at NASA's Kennedy Space Center (KSC). The suit-up room is the same room that has been used by all astronauts since the Apollo missions. Here, with the help of United Space Alliance (USA) technicians from the Crew Escape

Equipment Group (CEEG), the astronauts begin the lengthy process of pulling on their distinctive pumpkin suits.

Panel 11.1. Spacesuit

The spacesuits the astronauts wear during launch and landing are designed to hold communications equipment, oxygen tanks, parachutes, and enough water for a day. Rather than being designed to walk around and move like a spacewalking suit, the launch and entry suit is designed to keep the astronaut alive while they are seated. To achieve its function, the ensemble comprises several layers of thin clothing. The orange layer everyone sees as the astronauts walk out to the Astrovan on their way to the launch pad is simply the top layer, which actually comprises two layers. The inner layer is a rubber-like material similar to a wetsuit (this can be thought of as the balloon part of the suit because it is what holds the air if the suit has to inflate). The outer part of the suit is a flame-resistant Nomex material that is much stronger than the inner layer and gives the suit its shape. With the helmet, visor, and gloves locked in place, the suit fully encloses the astronaut in a cocoon inflated to 3.5 pounds per square inch, which is about the same air pressure as a person would find 30,000 feet above Earth.

To keep the astronaut alive after landing, the helmet is equipped with a valve that allows air in after the oxygen runs out. The suit is also equipped with a life raft and a bailing cup to scoop water out of the inside. Once in the water, the astronaut has various tools to help them alert rescue crews. For example, a set of flares are tucked inside one leg pocket and a survival radio is in another.

For the CEEG technicians (which includes six insertion technicians and 14 suit technicians), the work is even more demanding, since the suits have to be checked out several days in advance of a launch. For example, emergency oxygen bottles must be inspected and installed into the harnesses, the suit's liquid cooling system must be checked to make sure it's not leaking, the suit must be pressurized to check for suit leaks, and the technicians must perform an end-to-end check of the communication system.

By the time the astronauts arrive in the suit-up room, they know their suits have been thoroughly inspected and can start the first stage of donning. One of the first items of equipment the crew dons is the Maximum Absorbency Garment (MAG), an adult diaper designed to absorb the inevitable release of liquid while the astronauts lie on their backs waiting for launch. The first set of clothing the astronaut dons is a set of long-sleeve and long-pant thermal underwear lined with tubes, through which cooling water flows after the Advanced Crew Escape Suit (ACES) is donned. The astronaut then sits in a recliner and steps into the suit feet first. This involves a few gymnastic moves requiring the astronaut to first bend over before placing one hand in, followed by the other hand and wiggling their arms to try and push through the suit arms, hopefully getting their head somewhere near the neck ring in the process. Once this is achieved, the astronaut's back is sticking out, covered in blue underwear; the astronaut's legs and arms are in the suit, but the rest of the astronaut isn't! Next, the astronaut forces their head into the suit's metal neck ring, which would be relatively easy if it weren't for a tight neoprene dam that forms a seal around the astronaut's neck. The dam is designed to be the width of the crewmember's neck, so it's very tight when the astronaut tries to push their head through it. After some struggling, sweating (by the astronaut and the suit technician!), and feeling a little discombobulated, the astronaut is finally ensconced inside the suit.

The suit technician, after having checked everything looks correct, zips the suit closed and helps the astronaut with the next stage, which is donning the boots. After the boots, the helmet and gloves are locked into place with connecting metal rings, but these are only worn long enough for the suit technicians to check the ensemble for pressure leaks. Once the pressure check is completed, the helmet and gloves are removed until the astronaut is seated inside the spacecraft. Although the astronaut appears to be ready to make his/her way to the spacecraft, the suit technicians haven't finished. They pack the astronaut's pockets with survival gear such as flares and radios, and usually some personal items such as pens and flashlights. Finally, after their pockets are stuffed, the astronauts are properly suited up in their own private cocoon and ready to leave for the launch pad, but, before they board the bus, they must engage in a time-honored NASA ritual – a card game! The card game is an old NASA tradition called Possum's Fargo, a fighter pilot game in which the worst hand wins. Once the commander finally loses his hand, the crew exits the Operations and Checkout Building for the crew walkout (Figure 11.3) photo opportunity, after which they board NASA's silver Astrovan (Figure 11.4) for their ride to the launch pad.

Inside the Astrovan

Inside the van, the astronauts plug into cooling units located at each seat to avoid overheating in their bulky, heavy suits. Helping them during their short journey to the launch pad is an insertion technician, who also is a part of the closeout crew. In addition to assisting the astronauts into the spacecraft, the closeout crew also works as quality assurance inspectors before the launch. In addition to being the last set of eyes that inspect flight-critical items going into the spacecraft, the closeout team spend every day before launch making sure proper hardware is installed correctly, which can include anything from payload items to the solid rocket booster (SRB). On their way to the pad, the van passes several security check points where guards either salute or simply give a thumbs-up. Along the route, trucks are parked for their evacuation in the event of a launch pad explosion. As they approach the pad, the astronauts notice fire trucks and ambulances with crews clad in silver firefighting suits, but their attention is quickly diverted by the magnificent sight of Ares on the launch pad (Figure 11.5).

Figure 11.3. The crew exits the Operations and Checkout Building en route to the pad. Image courtesy: NASA.

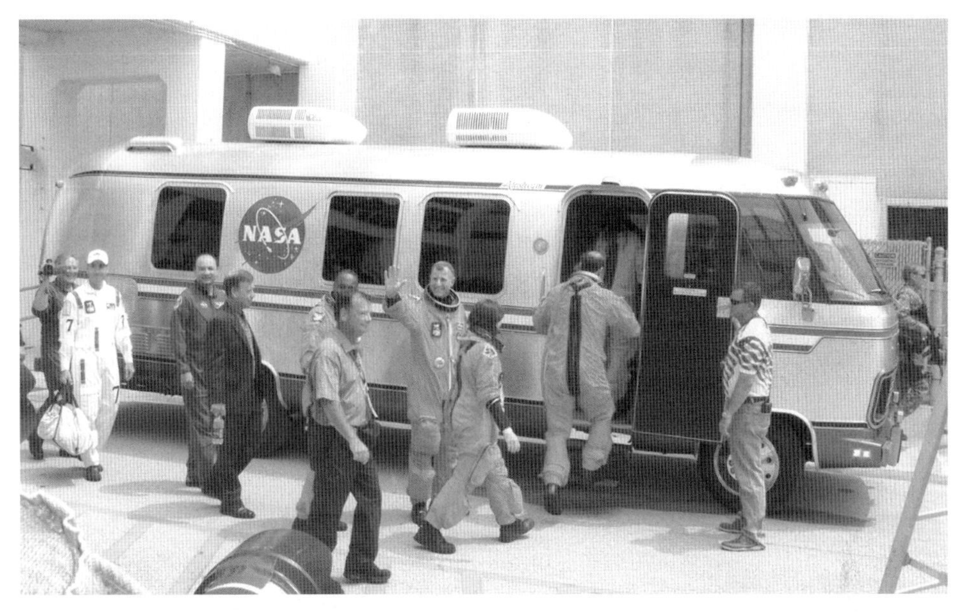

Figure 11.4. NASA's Astrovan. Image courtesy: NASA.

Figure 11.5. Ares on the launch pad. Image courtesy: NASA (*see colour section*).

White room

The process of assisting the astronauts into the spacecraft begins in the "white room" (Figure 11.6), where the closeout crew performs one final check of the suits, before strapping the astronauts into the spacecraft. Before being strapped in, however, the closeout crew helps each astronaut don a distinctive "snoopy cap", which contains the crewmember's communications headset. After this is done, the protective booties that covered the astronaut's boots during the trip to the launch pad are removed. Once the astronauts are secured inside the spacecraft, the closeout crew conducts tests on the communication systems, and checks oxygen lines and the myriad systems inside the spacecraft. While the closeout crew is doing this, the astronauts pull on their gloves and secure their helmets. Once they are happy everything is as it should be, the closeout crew removes all their equipment, says a goodbye to the crew, and closes the hatch. A moment later, the crew's ears pop as the capsule is pressurized. For the crew, the closeout crew may be the last humans they will see. The closing of the hatch is a sobering moment, since the closeout crew has trained with the astronauts for several months and a strong professional and personal bond has been formed.

Figure 11.6. The "white room". Image courtesy: NASA (*see colour section*).

Closeout crew

The hatch closing occurs about 30 min before launch. However, the job of the closeout crew is not over, because in the event of an emergency, they must go *back* to the launch pad to rescue the crew (while everyone else is evacuating the launch area!). For this reason, the job of a closeout team member is perhaps almost as dangerous as being an astronaut, since if they are required to rescue a crew (Panel 11.2), they are expected to enter an environment in which large quantities of volatile flammable fuels are present. For this reason, the closeout crew wears suits that are flame-retardant and anti-static.

Panel 11.2. Crew rescue

In the event the crew is incapacitated, the closeout crew has a really tough job. Imagine trying to extract a 70-kg astronaut wearing a flight suit weighing another 35 kg! Now imagine entering a dark, smoke-filled environment with water being sprayed everywhere, all the time knowing that at any minute, the whole launch pad could explode! For this reason, there is no question the closeout crew are some of the many unsung heroes of the space program.

If they are called upon in such an event, the closeout crew can identify the crew by means of a red glow light on the astronaut's arms (the closeout crew wear yellow lights for identification). Once extracted, the closeout team places the crew on the floor of the "white room", before putting them one by one in rescue chairs and rolling them over to the rollercoaster escape system. The unconscious astronauts would then be placed in the rollercoaster seats along with one closeout crewmember. The rollercoaster would be released and would slide (very fast!) to a landing area at the base of a bunker, where two fire-rescue personnel would be waiting. The astronauts would then be taken to the bunker and, in the event of a life-threatening injury, be placed in the M113 and evacuated from the launch pad.

Behind the scenes

The astronauts aren't the only ones dressed in bright orange protective suits who travel to the launch pad on launch day. The Final Inspection Team (FIT), also known as the "Ice Team", are an elite group who spend the better part of 2 hr inspecting a fully fuelled spacecraft as it stands ready for launch. It's definitely not your typical nine-to-five! The Ice Team has the potentially dangerous job of inspecting the entire stack to locate any unusual ice build-up or debris that might endanger the vehicle and crew during and after launch. If they notice anything out of the ordinary, they can make a call directly to the Launch Director.

Their day starts by suiting-up in protective gear, loading inspection equipment into their van, after which they receive a final safety briefing near the Vehicle Assembly Building (VAB). The safety briefing includes the very latest information that could help them exit the launch pad quickly in the event of an emergency. After the briefing, they wait for clearance from Launch Control to proceed through the security blockades to the pad. The signal from Launch Control is the T–3 hr mark in the countdown, at which point the Ice Team join the convoy of vehicles including the closeout crew and the astronauts. As they unload their equipment at the base of the launch pad, the team can hear the hum, creaks, and bangs emitted by the spacecraft towering overhead as the vehicle undergoes temperature changes. After unloading their equipment, they take the elevator to the top of the service structure, where they begin their inspection. Using cameras, binoculars, and infrared sensing devices, the team checks every surface of the vehicle, looking for unusual ice build-up and abnormal temperature readings that might indicate a problem. As the group continues working methodically at each level of the service structure, they regularly check in with Launch Control. They finish their inspection at the base of the vehicle on the mobile launch platform, moving around the SRBs' aft skirt. Following a final walk-around inspection of the vehicle, the Ice Team load their van and head back to Launch Control, where the team leader reports to the Launch Director.

LAUNCH COUNTDOWN

The countdown formally begins with the NTD issuing the call to stations from the LCC. The countdown clock begins ticking at T–43 hr about 3 days before liftoff (Table 11.1). With all the built-in holds, it takes approximately 72 hr to conduct a launch countdown. Guiding the NTD and Launch Director is a set of five thick manuals (reviewed and updated prior to each mission) documenting the procedures not only for the countdown, but also for the processing and testing of the spacecraft. The first manual describes the preparations leading to the beginning of the countdown, while the second manual includes the actual set of countdown instructions that sequentially configure the vehicle for launch. The third manual includes instructions to be followed in the event of a scrubbed launch; the fourth manual contains specific system instructions related to the second manual, while the fifth manual outlines preplanned contingency procedures and emergency instructions.

Table 11.1. Launch countdown milestones.

Time	*Event*	
T–HH	T–MM	Call to stations half an hour before countdown begins
43[1]	00	Countdown begins
27	00	Begin 4-hr built-in hold[2]
27	00	Clear blast area
27	00	Master events controller preflight built-in test equipment (BITE) test
27	00	Resume countdown
19	00	Begin 8-hr built-in hold[2]
19	00	Crew module clean and vacuum
19	00	Countdown resumes
19	00	Avionics checkout
11	00	Begin 13-hr 37-min hold[2]
11	00	Crew weather briefing
11	00	JSC flight control team on station
11	00	Communication activation
11	00	Crew module communication checks
11	00	Flight crew equipment late stow
11	00	Debris inspection
11	00	Ascent switch list
11	00	Resume countdown
10	40	Pad clear of non-essential personnel
09	50	Fuel cell activation
07	00	Red crew assembled
06	15	Fuel cell integrity checks complete
06	00	Begin 2-hr built-in hold[2] (T–6 hr)
06	00	Crew wake-up
06	00	Mission management team meeting
06	00	Resume countdown
03	00	Begin 2-hr 30-min built-in hold[2]
03	00	Closeout crew to white room
03	00	Astronaut support personnel communication checks
03	00	Pre-ingress switch configuration
03	00	Crew breakfast
03	00	NASA television coverage begins
03	00	Final crew weather briefing
03	00	Crew suit-up begins
03	00	Resume countdown
02	55	Crew departs Operations and Checkout Building
02	25	Crew ingress
01	35	Astronaut communication checks
01	20	Hatch closure
01	10	White room closeout
00	20	Begin 10-min built-in hold[2]
00	20	NASA Test Director briefing
00	20	Resume countdown
00	15	KSC area clear to launch

Table 11.1. *continued*

Time	*Event*	
00	09	Begin final built-in hold[2]
00	09	NASA Test Director launch status verification
00	09	Resume countdown
00	09	Start automatic ground launch sequencer
00	07	Retract access arm
00	05	Arm solid rocket booster and range safety safe and arm devices
00	05	Launch window opens[3]
00	02	Crew closes and locks their visors
00	01	Deactivate solid rocket booster joint heaters
00	00	Launch

[1] There are 43 hr of countdown clock hours, and not actual hours.
[2] There are a number of built-in holds in the countdown. These hold times are used to test systems, to communicate, and to cross-check.
[3] The point in time when the vehicle must be launched to meet specific mission objectives.

The instructions contained in the five volumes are conducted under the direction of the NTD, who starts the countdown by initiating a series of systems checks that progressively configure the vehicle for flight. During the countdown, there are several milestones, such as pad and vehicle closeouts, retraction of the rotating service structure (RSS) at the launch pad, activation of onboard fuel cells and, of course, boarding the flight crew!

Once the countdown clock starts ticking, the prime firing room is staffed around the clock and as countdown milestones are ticked off, assigned shifts of teams report on station at varying times, depending on which system they oversee. For example, the operators who oversee the ground launch sequencer (GLS) may be on hand earlier in the countdown, but they do not have a formal role until the day of launch. This crew, comprising a primary operator, two backups, and a fourth to retrieve data, is on station by the T–2 hr mark to call up the GLS software and prepare for the final count. At T–20 min, the GLS begins issuing active commands, and at T–9 min, it assumes automatic control of the count.

Launch commit criteria

As the various operators work through the endless sequences comprising the countdown, the launch team members at the consoles in the prime firing room are monitoring vehicle and support system performance against predetermined parameters. This process is particularly lengthy, as the number of checks that must be made runs into the thousands, about two-thirds of which originate from the flight vehicle and one-third from the ground support equipment. The checks usually fall into the category of Launch Commit Criteria or supporting data. Launch Commit Criteria are those factors that have implications for safety and/or mission success.

Basically, they define the conditions under which a vehicle can and cannot launch. The Launch Commit Criteria checks begin at T–6 hr and include parameters guarding against flight hardware damage and those designed specifically for astronaut safety. For example, there are certain weather considerations that must meet specific Launch Commit Criteria requirements before a launch can proceed.

The other set of data comprises supporting information available for the engineers to help maintain a specific hardware configuration or to aid in troubleshooting problems. For example, there are certain temperature limits for vent lines carrying hydrogen to the vehicle. The supporting information is part of a huge database of system performance that, in turn, is part of the Launch Processing System (LPS).

Communication

With so much information being exchanged and processed, controlling the communication traffic is essential. In the prime firing room, communication is routed through the Operational Intercommunications System (OIS), a closed-loop digital voice system utilizing fiber-optic cable. During countdown, the NTD uses one frequency as the command channel, while the Test Conductors use separate channels to individually lead their part of the countdown. The Test Conductors and the NTD communicate with each other as the situation demands on issues such as status checks and command responses. The use of different channels ensures distinct lines of communication and keeps traffic to a manageable level. Outsiders listening in to the communication might be forgiven for thinking the launch team is speaking a foreign language due to the heavy use of acronyms (like this book!) – a tactic used liberally by the team to keep verbiage to an absolute minimum. Verbiage is reduced even more by the unique call signs assigned to firing room console positions – a strategy used by the team for quick and positive identification.

The only time communication channels change during the countdown is once the crew enter the spacecraft, at which point the astronauts hook up to the Test Conductor's channel. At the T–20 min mark, the NTD also switches to this channel and it becomes the Command channel. Meanwhile, the integration console in the prime and backup firing rooms maintains communication channels with the other centers, such as the Mission Evaluation Room (MER) at JSC and the Huntsville Operations Support Center (HOSC) at the Marshall Space Flight Center (MSFC). The MER plans and implements flight data retrieval, processing, evaluation and reporting, and post-mission evaluation while the HOSC provides technical support on the vehicle.

Firing room

The military-type discipline exercised in the prime firing room is a reflection of the critical decisions taken by those charged with the lives of up to half a dozen astronauts. Those working in this high-pressure environment receive special training in the particular rules and regulations governing their conduct. For example, conversations must be limited to the business at hand, no personal telephone calls are permitted except in a dire emergency, and no one is allowed to read non-work-related materials. From the T–3 hr mark, entrance into the prime firing room is restricted to personnel with firing room badges. At T–20 min, the door to the prime firing room is locked, the intent being to eliminate distractions and allow the team to focus its attention on the countdown.

Pad activities during countdown

While countdown activities are being controlled from the prime firing room, personnel at the pad perform different tasks required for launch preparations. Approximately 15 min before launch, readiness polls are conducted by the three teams comprising the launch team. First, the NTD verifies the prime launch team reports no violation of Launch Commit Criteria. Second, the Engineering Director, who heads up the Engineering Support Team, verifies no constraints for continuing the final count. Finally, the Mission Management Team Chairman verifies there are no issues with any of the senior element managers. These three verifications are then passed on to the Launch Director, who conducts a KSC management poll. Assuming all the key personnel agree, the Launch Director gives his permission to proceed with the countdown to the NTD and the NTD sets in motion the final 9 min of the countdown, automatically controlled by the Ground Launch Sequencer (GLS).*

Vaya con Díos**

Inside the capsule, the crew performs radio checks, follows the seemingly endless sequence of countdown milestones, and listens to the LCC dialog. One of the call signs that usually gets attention is the Range Safety Officer (RSO). The RSO (Panel 11.3) has the onerous job of transmitting the Flight Termination System (FTS) that destroys the vehicle in the event it strays off course following launch. If such an event were to occur, the commander would be warned by a red light illuminating on the instrument panel. From a human factors design perspective, the light isn't very helpful!

---

* The GLS is an automated program which controls all activity during the final portion of the countdown.

** "Go with God", in Spanish.

Panel 11.3. Range Safety Officer

Ares will be launched into a space above the launch range called the *launch corridor*. If the solid rocket booster (SRB) fails while the Ares flies inside the launch corridor, the rocket will fall in an uninhabited area. Failure outside the launch corridor, however, may cause Ares to fall on people or property. In such an event, the RSO has authority to order the remote destruction of the launch vehicle. Such an event occurred following the disintegration of the Space Shuttle, *Challenger*, when the RSO ordered the uncontrolled, free-flying SRBs destroyed before they could pose a threat.

Destroying a launch vehicle is relatively easy. Ares has a Range Safety System in its SRB capable of receiving two command messages – *arm* and *fire* – which may be transmitted by the RSO from the ground station. The first message, called *arm*, enables onboard logic to enable a destruct and illuminates a light on the flight deck display and control panel at the commander and pilot station. The second message that may appear is the *fire* command.

Trying to forget about the RSO, the crew continues to monitor the checklist, mentally ticking off events in their head and praying there won't be any unscheduled holds. Due to the complexity of the launch sequence and the myriad vehicle systems that must function perfectly before the go for launch is given, it is not surprising many launches are delayed by one problem or another. These problems may include anything from failure of a computer system to light aircraft entering the restricted airspace around the launch pad. Unsurprisingly, for astronauts looking forward to their first flight after years of preparation and training, the problems do little to soothe the nerves. At this point in the count, the rookies are anticipating an event they have dreamed about since early childhood and can't wait for the hold-down bolts to blow, releasing the vehicle on its 9-min ride into space. The mindset of a rookie astronaut just minutes from launch is similar to that of a mountaineer approaching the summit of Mount Everest. While the odds of returning from the summit of Mount Everest (about one in 12) are a little better than manned spaceflight (about one in 65), the term *summit fever* applies equally to potential space farers. Astronauts have their own quest and, in common with high-altitude climbers, are motivated by a fear far greater than death – the fear of not reaching the top. This fear is partly manifested by astronauts reciting the astronaut prayer:

"Please, dear God: Don't let me screw up."

Alan Shepard, the first American in space, shortly before launch on May 5th, 1961. Shepard's pre-launch quote has since become known among aviators as *Shepard's Prayer*, and among astronauts as the *Astronaut's Prayer*

T–10 minutes

As the clock shows the T–10 min mark, the astronauts check their straps one more time, listen to the endless series of acronyms being checked off by the flight controllers, and watch the commander follow the checklist. In just a few minutes, the vehicle will lift off from the pad in a thunderous cloud of fire and smoke.

T–2 minutes

With 2 min before launch, the MCC reminds the crew to close their visors, a command confirming launch is imminent. The crew oblige, snapping the bail into place to seal the faceplate. On their left knees, the crew flips a lever on their suits to start the flow of *Orion*'s oxygen. With adrenaline pumping and their heart rate close to the red line in anticipation, the crew tries to slow their breathing and control the attack of butterflies. For a moment, the thought that these could be the final moments of their lives enters their head, but their attention is quickly drawn to the expectancy of Ares lighting up. After 20 years' imagining riding a rocket into space, after thousands of hours of training, after all the sacrifices, the rookies are finally going to get their chance at wish fulfillment. Bodies tingling with anticipation, the rookies give each other a thumbs-up.

T–1 minute

With 1 min to launch, the PFDs continue scrolling data as the astronauts think about their families and all the training and preparation that brought them to this pivotal moment in their lives. The astronauts' families are 3 miles away, standing on the roof of the Launch Control Center (LCC), and are probably more scared than any of the astronauts. It might be an Ares rocket standing on the pad, but in the mind's eye of the families, the image of *Challenger* blowing up 73 sec after launch is still a powerful one. The MCC announces 10 sec to launch – a command indicating the GLS has taken over the count. Next is the command "GLS is Go for auto sequence start". Unlike the Space Shuttle, which started its noisy liquid-fuelled engines 6 sec before launch, there is nothing inside *Orion* to suggest the launch is imminent. At 15 sec to go, the attitude indicators on the PFD scroll to the correct launch attitude as the vehicle processes its final update.

T–5 seconds

At T–5 sec, tonnes of water flood the pad and cascade into the flame trench designed to absorb the shock waves of ignition. Finally, the countdown reaches the endpoint. Ignition commands from the vehicle's flight computers commit the vehicle to flight as electrical energy is routed from the first-stage avionics to the motor igniter. The igniter then initiates the burn of the five-segment motor. As the motor burns, it builds up combustion along the surface of the propellant, which is expelled out of the nozzle, creating the thrust for lift-off. The pyrotechnic charges holding the vehicle to the pad explode and the vehicle shoots toward the sky, releasing 3.5 million pounds of thrust. A rumble shakes the stack from 12 stories below as the main engine spews fire and roars to full power. On the PFDs, the computers scroll rapidly through engine checks and mark the mission time: zero. The clock is running! Inside *Orion*, the astronauts are pushed back into their seats by a force more than 1.5 times the force of gravity. The MCC makes the call: "Tower Clear!" The SRB is already consuming propellant voraciously at 5 tonnes/sec, burning at a searing 2,760°C. The crew are pushed further back into their seats as the gravitational forces ramp up to nearly 3 G. On the roof of the LCC, the families watch awestruck as Ares blazes a trail of flame (Figure 11.7) against the blue sky. As the thunder of the blazing SRB reverberates around the LCC, shaking the roof, some family members snap pictures while others simply stand there, stunned by the spectacle of the unrestrained fiery power being unleashed just a few miles away.

L+

Less than 30 sec after launch, shock waves form on the nose and the vehicle begins to vibrate. Meanwhile, the vehicle's control system continuously monitors trajectory and issues guidance commands to ensure the rocket nozzle's angle is correct. A few seconds later, the commander announces Mach 1 and the crew listen to the rushing sound of supersonic air being displaced by the vehicle. Quickly, the sky turns from blue to black. A few seconds later, a loud bang shakes *Orion* as the SRB is released and falls away. A parachute will lower it into the ocean, where a recovery ship will pick it up so it can be used again. As the SRB falls away, the cockpit becomes more silent and the ride becomes noticeably smoother. The commander informs the crew the SRB is spent, a signal for them to shut off the oxygen and open their visors. The vehicle is now more than 2 min into flight and flying at an altitude of more than 60 km. The astronauts feel another thump as the upper stage's J-2X engine ignites, propelling the vehicle to Earth orbit. A few more seconds pass and the commander announces an altitude of 100 km. The rookies let out a cheer, since the altitude means they are now officially astronauts and have collectively achieved their life goal. Meanwhile, Ares, reduced to its first stage and the capsule, continues to accelerate, the altimeter tape scrolling madly on the PFD. Mach 15. Ares is tearing along at more than 5 km/sec! Mach 18 – 6 km/sec! With a force of more than 3 G pressing down on them, each crewmember weighs more than 300 kg. Twenty times the speed of sound! And still it continues to accelerate. Finally, the Mach meter reaches Mach 25 – a velocity of more than 8 km/sec and the commander welcomes the crew to space. The elation of having realized a dream begins to sweep over the rookie astronauts, who are no longer rookies! One of them asks the commander if they can do it again. It's a beautiful moment.

Figure 11.7. Launch of *Ares I*. Image courtesy: NASA (*see colour section*).

Index

Printing: Mercedes-Druck, Berlin
Binding: Stein+Lehmann, Berlin